IMAGES of America

MARINELAND

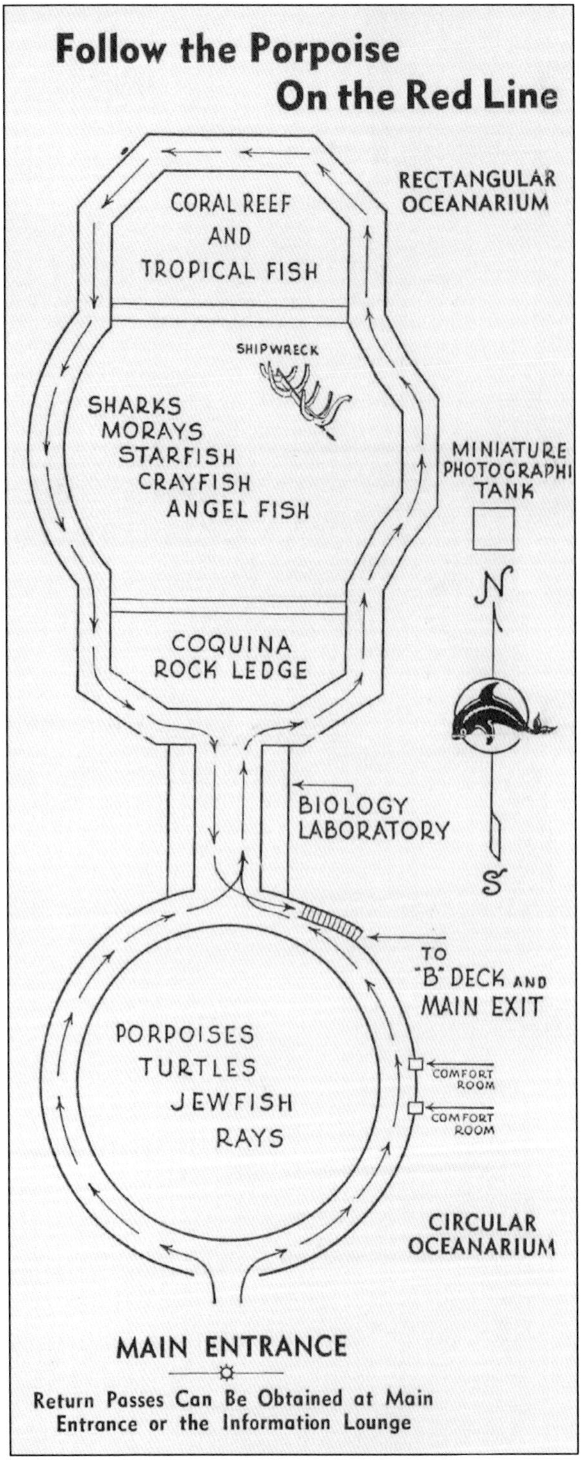

This map depicts the original layout of Marine Studios, the world's first oceanarium, as it stood on opening day, June 23, 1938. This project was the dream of several innovative men who imagined a facility where different species could live together as they do in the sea rather than segregated, as they were in traditional aquaria of the day. (Courtesy of Marineland archives.)

On the Cover: Jumpmaster Mitch Lightsey performs for the anxious crowd with Dottie, an enthusiastic spotted dolphin, one of the many rare and unusual species to be displayed at Marineland of Florida over its 70-year history. (Courtesy of Marineland archives.)

IMAGES
of America

MARINELAND

Cheryl Messinger
and Terran McGinnis

ARCADIA
PUBLISHING

ISBN 978-0-7385-8220-7

Published by Arcadia Publishing
Charleston, South Carolina

Printed in the United States of America

Library of Congress Control Number: 2010939085

For all general information, please contact Arcadia Publishing:
Telephone 843-853-2070
Fax 843-853-0044
E-mail sales@arcadiapublishing.com
For customer service and orders:
Toll-Free 1-888-313-2665

Visit us on the Internet at www.arcadiapublishing.com

This book is dedicated to the remarkable people and animals of Marineland, past, present, and future.

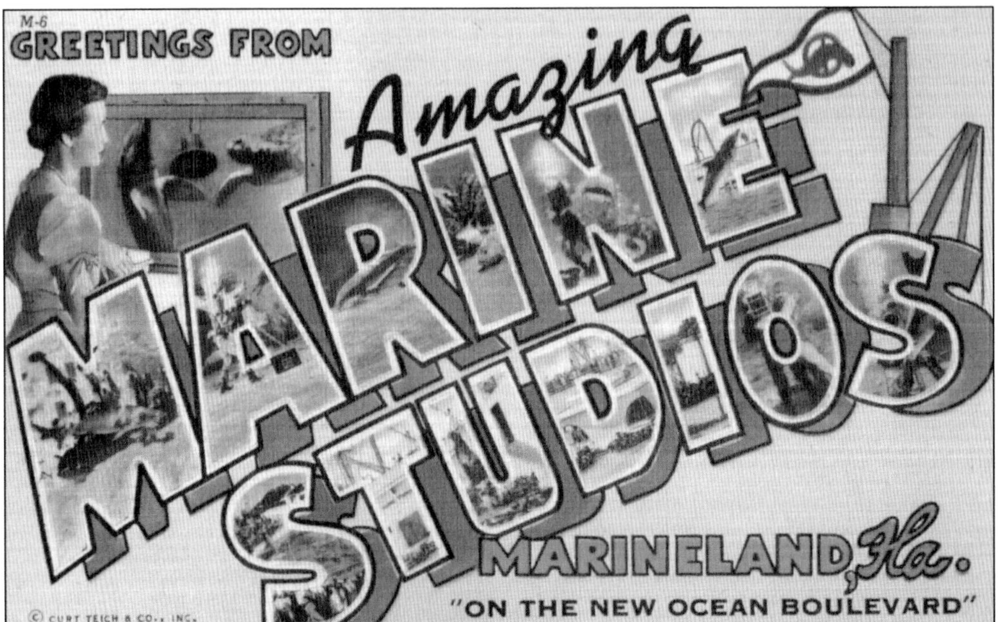

This early postcard was a souvenir for visitors to send back to their friends and family who had undoubtedly never seen or even heard of an oceanarium. (Courtesy of Marineland archives.)

Contents

ACKNOWLEDGMENTS

We are very grateful for those individuals who, for generations, saw the value in maintaining Marineland's irreplaceable archive. Their allocation of time and resources to the preservation and maintenance of the documents, records, and photographs, perhaps knowing that someday there would be a story to tell, has made our job so much easier. Our sincere gratitude is extended to the current Marineland owners, management, and staff for entrusting us with unlimited access to this priceless collection. All images are courtesy of the Marineland archives. We have poured heart and soul into this book in an attempt to provide an accurate telling of Marineland's amazing story, and any mistakes are purely our own. We would be remiss without a very sincere and personal note of thanks to Dr. Sam Ridgway, Marilee Menard, Jeff Jouett, Billy Hurley, Doug Messinger, and Kevin McGinnis, who each inspired, encouraged, and supported us in this work. Our hope is that this book sufficiently honors the people and animals whose lives have been so influential in creating the world's first oceanarium.

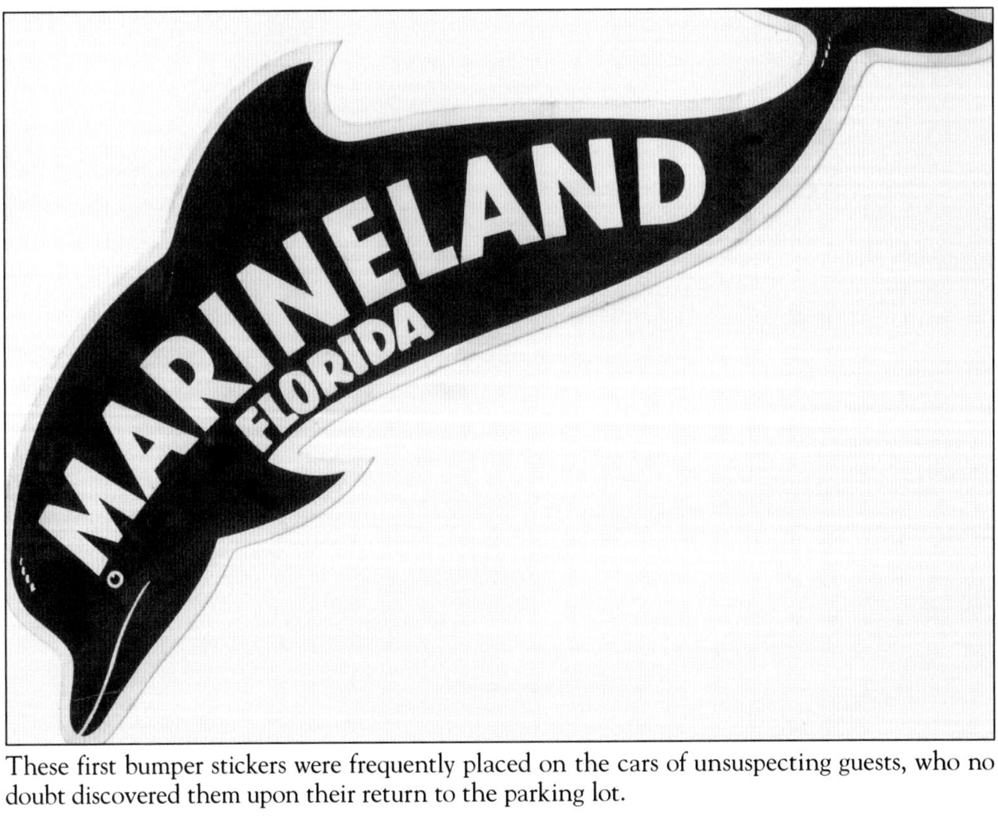

These first bumper stickers were frequently placed on the cars of unsuspecting guests, who no doubt discovered them upon their return to the parking lot.

Introduction

Life was simple in the 1930s. The Great Depression left Americans with little money and a renewed resolve to enjoy the simple things in life—dancing to the sound of big bands, listening to weekly radio programs and Roosevelt's fireside chats, playing board games in the parlor, cheering for Babe Ruth and the Yankees, and escaping reality with Agatha Christie mystery novels and Hollywood movies at the local picture show. During this wholesome era in American history, a group of visionary men appeared. These daring men's lifestyles of exploration, adventure, and filmmaking (made possible by their ties to some of America's greatest fortunes) were larger than life, yet they had more than their fortunes in common. They also shared a dream that was inspired by their adventurous spirits—to make the real life drama of the undersea world accessible so that it could be witnessed, studied, and filmed. Certainly aquariums existed, but none like this. With a pool large enough to realistically mimic the ocean, all marine life, including fish, sharks, sea turtles, whales, and dolphins, could live together as they would in the wild, unsegregated and untamed. The display was so unique that it prompted the coining of a new name to describe it: oceanarium, the ocean in an aquarium. In an era when only a handful of the strongest and bravest men were hard-hat deep-sea divers, the world's first underwater film studio was created, aptly named Marine Studios, introducing the ocean to the world in a novel and fascinating way. Although Marine Studios provided exceptional opportunities for filming, it was the scientific and educational value of this endeavor that captivated the world.

One of the most important decisions was choosing the right location. The original concept for Marine Studios had been a natural ocean pen on the California coast, a project that was quickly abandoned due to the uncontrollable nature of the natural sea environment. Opting instead to build a pool where the water quality parameters could be controlled, placement of the studios was critical. It had to be in close proximity to coastal marine environments, which would allow access to fresh seawater, as well as emerging tourism activity, and Florida was just the place. Following years of searching, a spit of land was selected along coastal Highway A1A, just 18 miles south of St. Augustine. It was flanked on the east side by the Atlantic Ocean, while to the north there was easy passage through the Matanzas Inlet to the calmer waters of the Intracoastal Waterway in the west, a perfect location for boat dockage, which meant easy access for obtaining specimens. In addition, its proximity to the Gulf Stream not only afforded accessibility to clean, clear water, but also provided relative protection from hurricanes.

With the site secured, construction began in May 1937. And although this was a pioneering endeavor, it took only one year to complete construction, as well as fill and stock the pools with a multitude of creatures from the sea. The world's first oceanarium opened to the public on June 23, 1938, and quickly received accolades, becoming the state of Florida's most successful and popular tourist attraction. Its special allure attracted a seemingly nonstop stream of scientists, writers, artists, celebrities, and members of the media. For the first time, showmanship was added to the display of marine life, thanks to daring deep-sea divers and friendly sailor-clad dolphin trainers. But the truth of the matter is that it was all made possible by the work conducted behind the scenes. There were few books to read, journals to reference, and colleagues to call who had experience in the exhibition of mixed species. For this reason alone, the scientists and researchers of Marine

Studios were extremely innovative. They made hundreds of discoveries and often invented solutions to the unique challenges they faced. Every day presented a new opportunity to learn.

The founders of Marine Studios were wise businessmen who developed not only the oceanarium, but built an entire town to provide a pleasurable, all-inclusive retreat for visiting guests. Along the adjacent beachfront property, they featured a hotel with outdoor bowling, shuffleboard, tennis and ping-pong; a bathhouse complete with dressing rooms and showers; and a full-time athletic director. Refreshments were served at the Penguin Bar, the Dolphin Restaurant, and the Periwinkle Coffee Shop. There was a full-service gas station, a Greyhound bus station, and a gift shop that featured unique ocean-related gifts, Florida novelties, postcards, souvenirs, and express shipment of fresh Florida citrus. For those guests who wished to arrive by boat, there was a yacht dock and fishing camp, complete with a bait-and-tackle store, fishing guide service, and a dockmaster on duty 24 hours a day. But, by far, the most famous attraction, besides Marine Studios itself, was the Moby Dick lounge, where guests could enjoy their cocktails aboard a bar that rocked like a true ship. As a community set up to entertain visitors, the city of Marineland was truly becoming one of Florida's newest and most complete vacation destinations.

Even today, there remains much confusion over the use of the seemingly interchangeable names Marine Studios and Marineland. The land adjacent to the Marine Studios attraction became the town of Marineland, Florida. Perhaps the smallest official town in the state, with a population of fewer than 10 total inhabitants, it was officially incorporated 12 years after the opening of Marine Studios on November 1, 1946; however, the name of the town Marineland became synonymous with the attraction Marine Studios. So, in a classic turn of events, Marine Studios officially changed its name to Marineland of Florida in 1961. Moreover, the founders had never really abandoned their original vision of settling in California. In fact, they lent their expertise to open another major oceanarium, Marineland of the Pacific, along the California coast in 1954.

In the 1970s, after its first 40 years, the ever-innovative and ever-adapting Marineland underwent its largest expansion, complete with additional exhibits featuring penguins, sea lions, otters, river dolphins, and electric eels as well as a 3-D movie theater. Although trained dolphin shows had been at the heart of Marineland's popularity, there was a trend developing toward increased participation and less passive audiences by the late 1980s. This trend, combined with a push from Mother Nature in the form of several damaging hurricanes, led to the construction of a brand new facility in the 2000s, Marineland's Dolphin Conservation Center, designed specifically to allow guests to meet the dolphins up close and personal.

The founders of Marineland had amazing vision. They hoped to create a place of sound and lasting value for the public, the zoo and aquarium community, and the scientific world. Through stories and photographs, this book celebrates more than 70 years of pioneering history in marine science, education, animal care, and welfare. Get ready for a fascinating journey.

One

Pioneers

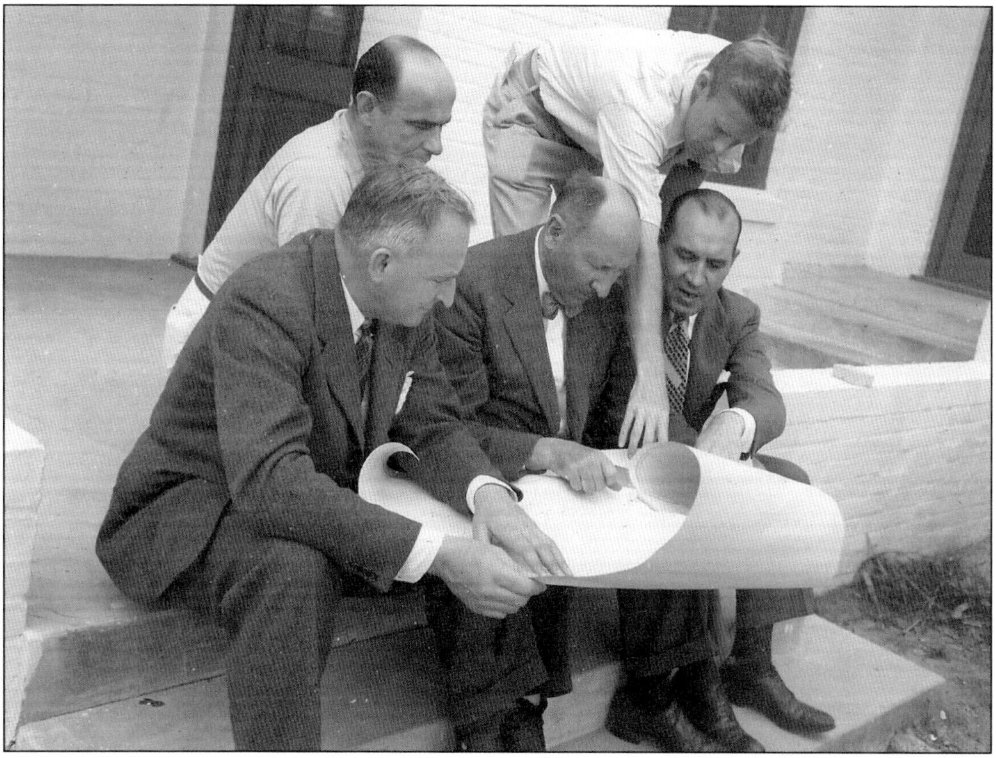

It was truly a pioneering spirit that motivated a group of naturalists and businessmen to create the world's first oceanarium. Leaders in their own right, they brought together a wealth of experience to achieve feats in engineering and construction that had never before been attempted. Equipped with desire and a dream, they prevailed over each new challenge, transforming their vision into reality.

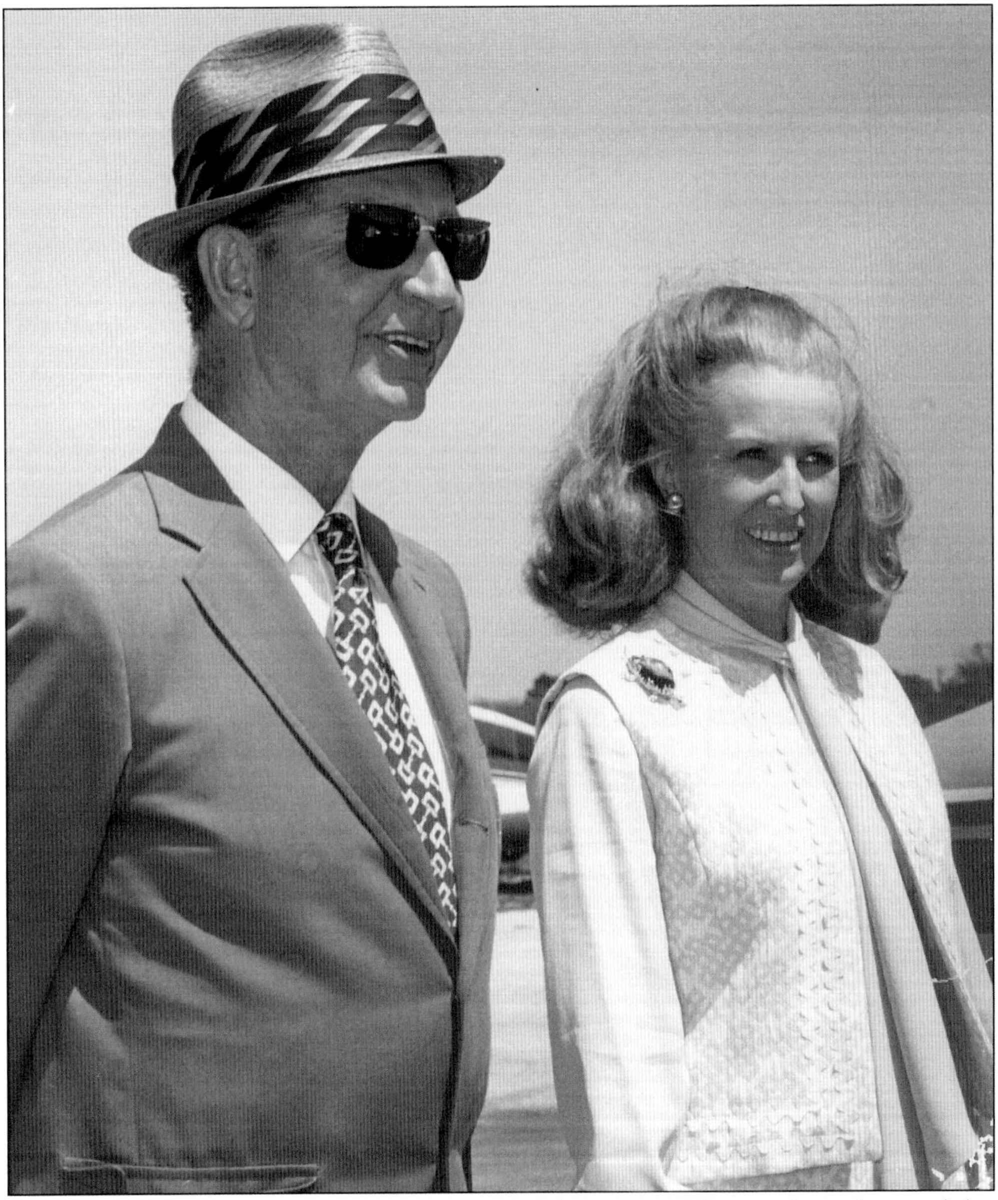

As a founder of Marine Studios, Cornelius Vanderbilt "Sonny" Whitney, shown here with his wife, Marylou Whitney, was a very capable businessman with involvement in early aviation, the motion picture industry, mining, museums, and horse racing. Whitney, who was a graduate of Yale University and an original founder of Pan American Airlines, proved himself to be successful beyond just his inherited name and wealth. A supporter of the arts and education, Whitney was a trustee of the American Museum of Natural History and a financial supporter for the making of *Gone With The Wind*. Whitney's involvement with Marine Studios extended beyond his interest in successful businesses or film production. His dedication to science and discovery was evident from the beginning in the prioritization of research at Marine Studios and in his donation of land and capital for the development of the marine science laboratory, named the Whitney Laboratory for Marine Bioscience, to the University of Florida. Whitney was the chairman of the board of Marine Studios at the time of its opening.

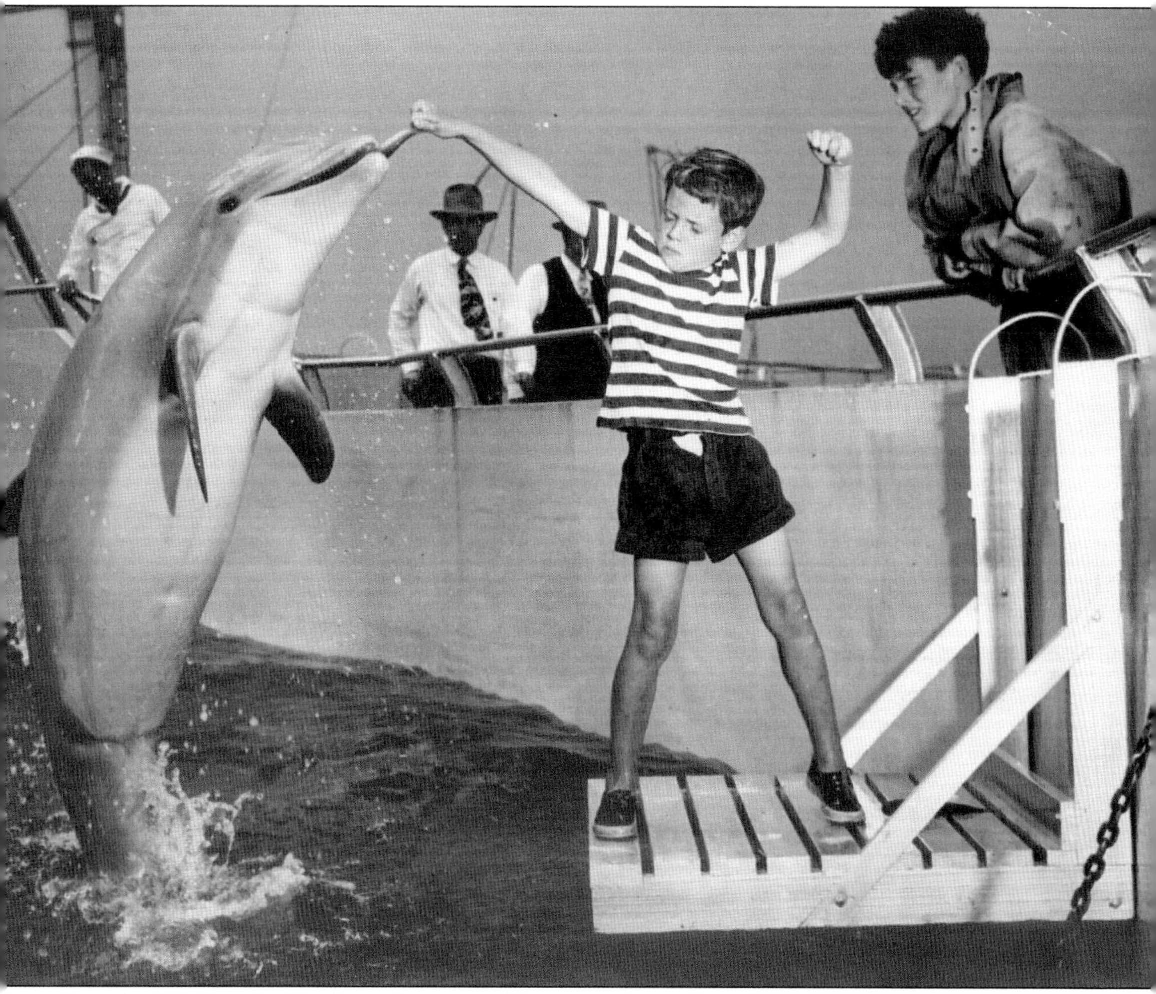

Founder W. Douglas Burden, whose son Christy is shown here feeding a dolphin at Marine Studios in 1948, had a lifelong passion for nature and discovery. As the great-great-grandson of shipping and railroad tycoon Commodore Cornelius Vanderbilt, he was born into a life of privilege. Long before his involvement with the Marine Studios project, Burden was a naturalist who enjoyed traveling and wildlife study. As a Harvard graduate and member of the American Museum of Natural History's board of trustees, he embarked on an expedition to Komodo in 1926, being the first to collect live Komodo dragons from Indonesia, two of which were displayed for the first time at the Bronx Zoo. This adventure is said to have been the basis for the movie *King Kong*. Burden wrote about this experience in his book *Dragon Lizards of Komodo*. After studying animals in the jungles, Burden wondered whether there might be equally dynamic behavior to be observed in the oceans. It was then that the idea for Marine Studios, of which he would become president, was born.

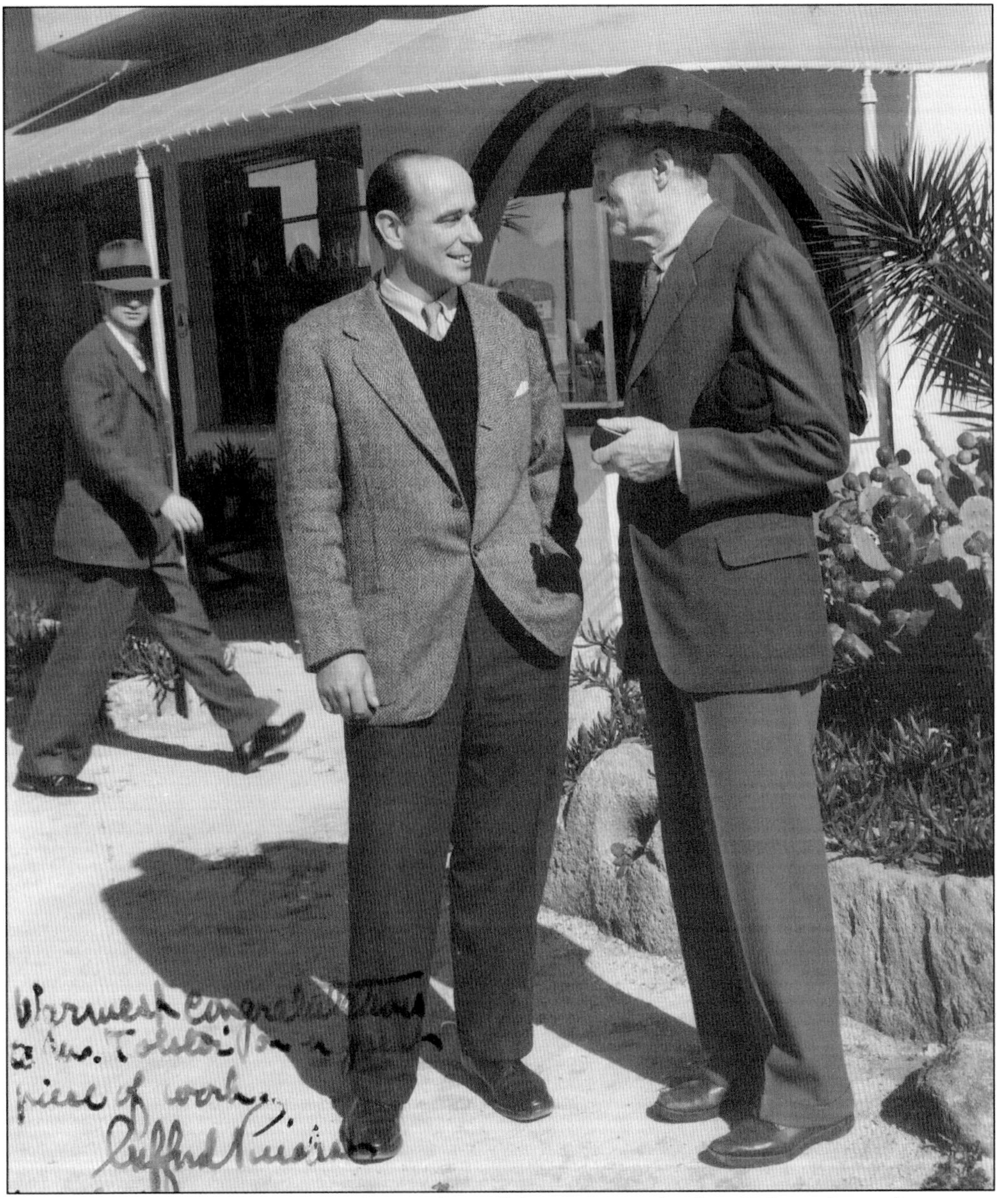

Founder Count Ilia Tolstoy (center), grandson of the great novelist Leo Tolstoy, began his career training horses in Russia. At only 12 years of age, Tolstoy became the assistant to the director of Cavalry Training Command. After escaping the Bolshevik reign in 1923, he studied animal husbandry at Iowa State College. When faced with the Great Depression, Tolstoy survived by training horses and offering riding lessons. A real turning point came when Tolstoy partnered with Otis Barton, the inventor of the bathysphere, an early submarine, to develop the first motor-driven underwater camera, which they used to make a documentary film on the sponge industry in the West Indies. These inventions were crucial steps toward making undersea life visible to the rest of the world, and Tolstoy's fascination was born. Throughout his life, Tolstoy had a strong dedication to conservation, including involvement with the Bahamas National Trust and the Caribbean Conservation Commission. In 1940, Tolstoy obtained U.S. citizenship and renounced his title of count. Tolstoy was the vice president of Marine Studios on opening day.

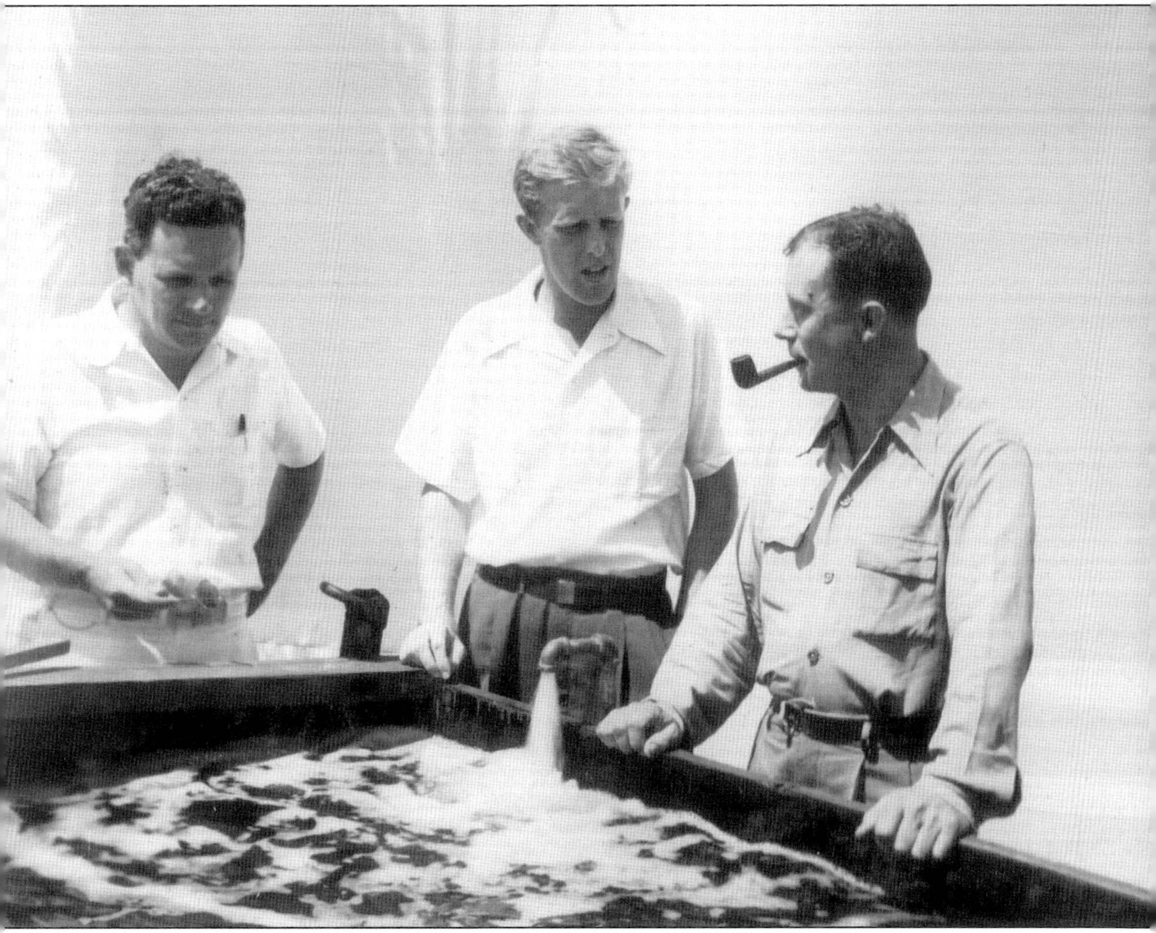

Although frequently credited with the development of Marine Studios, Whitney, Burden, and Tolstoy were not alone. Fellow founder Sherman Pratt had ties to RKO Pictures and was an active member of the Explorers Club, which promoted and supported scientific exploration and education. Pratt came from a long line of entrepreneurs; his father was a founder of the Boy Scouts, a trustee of the Museum of Natural History, and the conservation commissioner of New York, while his grandfather partnered with John D. Rockefeller to create Standard Oil. Pratt served as the motion picture consultant for Marine Studios, but the success of day-to-day operations must be credited to Arthur McBride, who is seen at center in the image above. McBride was involved in the collection, care, and study of the early animal collection as well as maintaining water quality and clarity. Much of our early understanding of dolphin behavior, reproduction, and anatomy can be attributed to the young McBride who was hired as the first curator of Marine Studios at only 23 years of age.

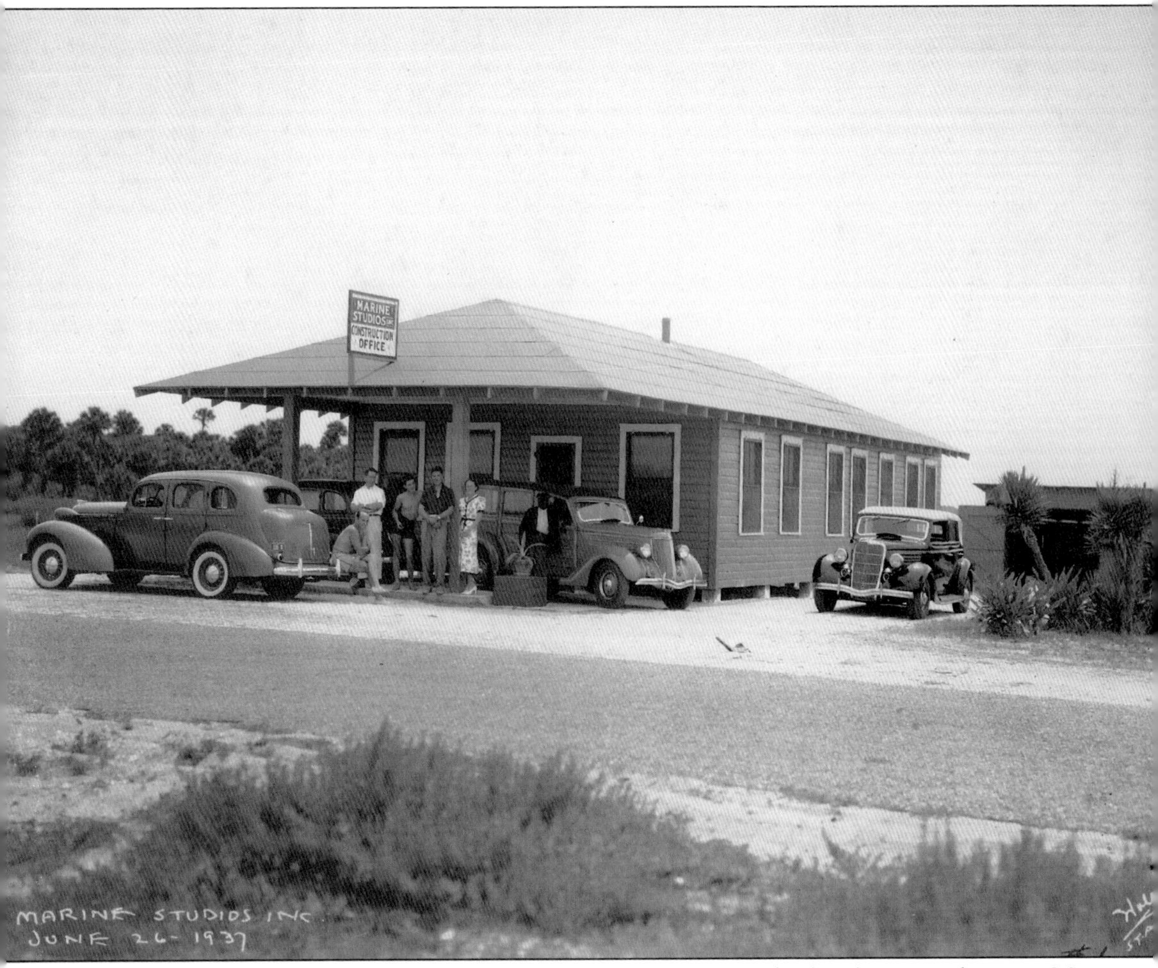

The Marine Studios construction office, shown here just one month after the project began in May 1937, was a retired gas station that had served for years as a communication point for rumrunners. It was here that much discussion and deliberation took place. With no precedent for building a facility like this, nothing could be done quickly or without forethought. Architects and engineers, scientists and film producers—everyone had an opinion on the design and a stake in its outcome. The engineer was M.F. Hasbrouk, who had constructed swimming pools at the White House and the Roosevelt family home. The lead architect, Fred Henderich, was from St. Augustine, where he pioneered the city's restoration project. He was joined by architect Jack Wood of New York. Amazingly, Marine Studios cost only $500,000 and was completed in February 1938, after only nine months of construction.

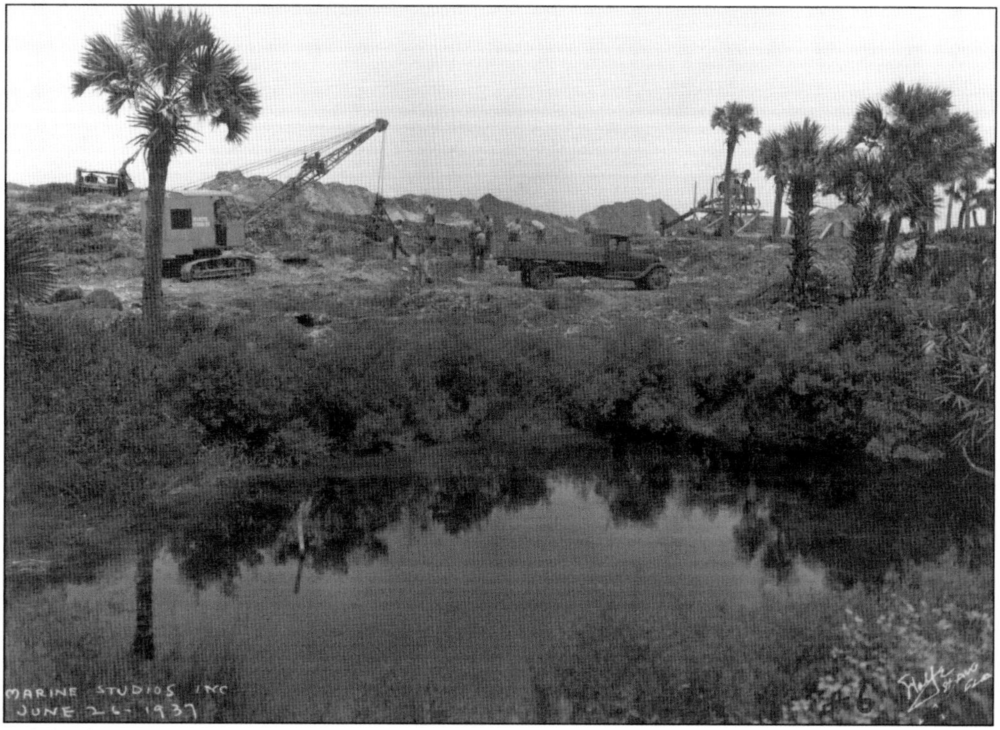

While the owners and investors were eager to begin construction, some early discoveries gave them reason to worry. Once committed to the site, they became concerned about area tourist traffic patterns and paid a young local boy to count cars for one week as they drove by the site on Highway A1A. The most he counted were six in one day. Another event that gave reason for pause was the discovery of an ancient burial ground of the Timucua people, an extinct tribe of Native Americans, near the construction site. Despite these situations, they forged ahead, and 300 tons of steel was soon delivered on barges via the Intracoastal Waterway, and the structure began to take shape along the barren coastline.

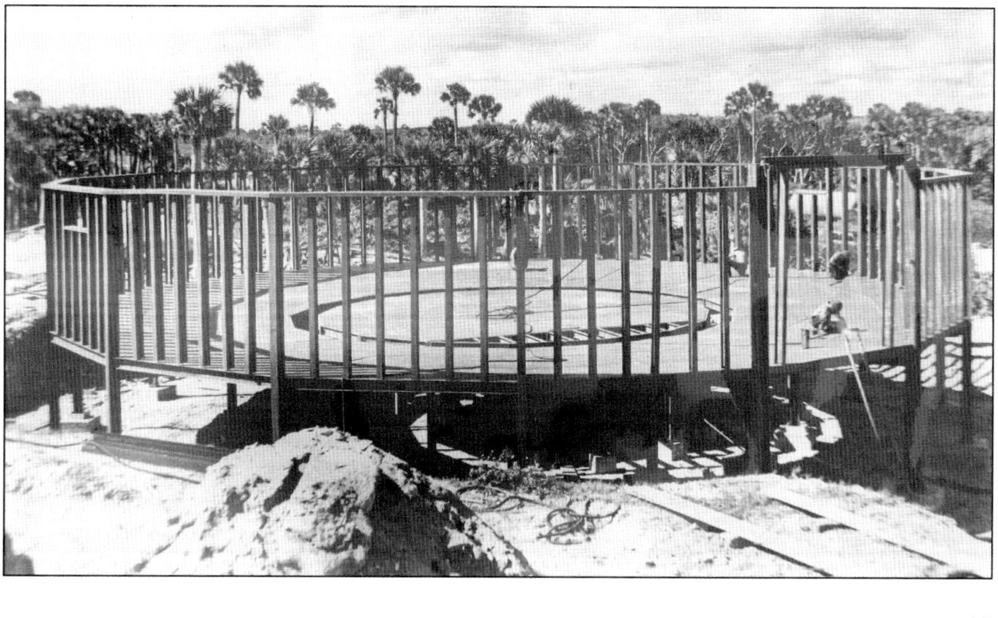

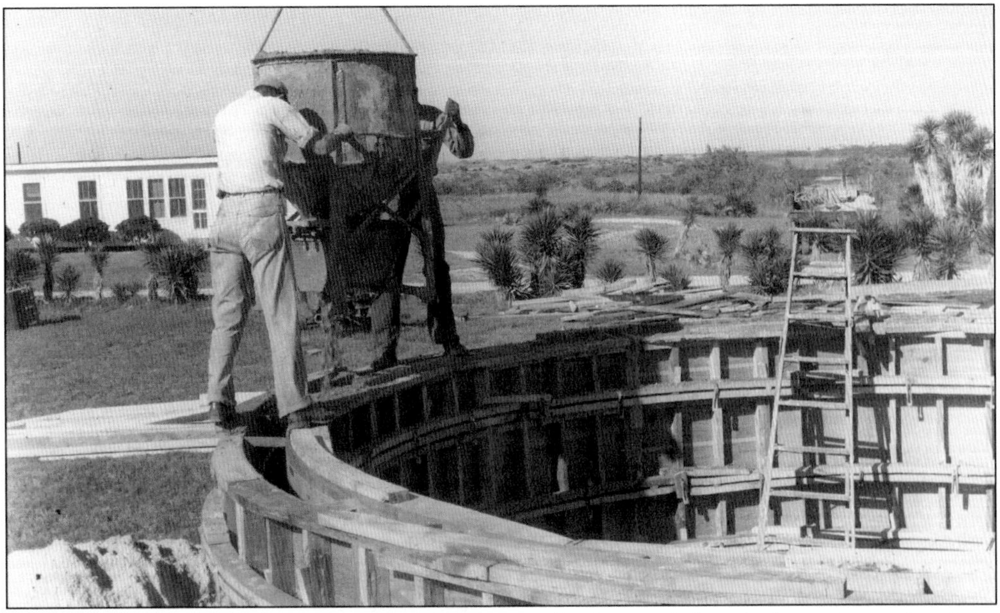

With the land up and down the coast almost completely uninhabited, the noise and activity being generated at the Marine Studios construction site seemed incongruous and contrary to the ultimate atmosphere desired. The extremes began to take their toll on the workers. Along with occasional bouts of frustration and pessimism about this seemingly impossible construction task came physical discomfort from working outside in severe weather conditions, along with feelings of isolation resulting from the lack of phone lines to the area. Despite the difficulties, work progressed while Fred Waller of Courier Productions worked out the technical details for the placement of the 200 portholes to allow for the best camera angles. As the walls rose like nothing ever seen before, the exhausted construction crew began to feel they were part of something historically groundbreaking.

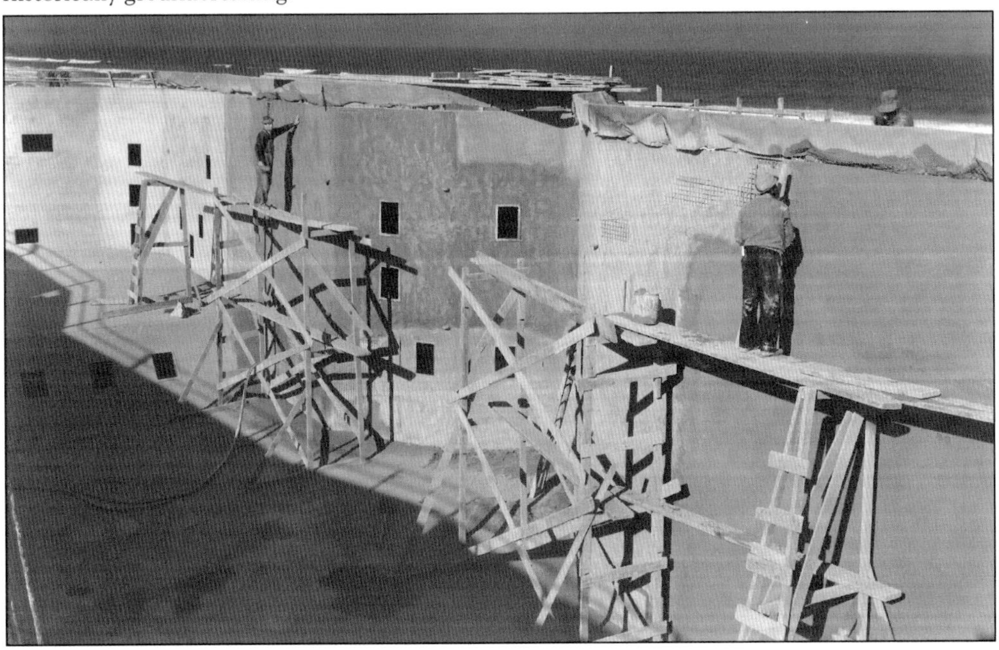

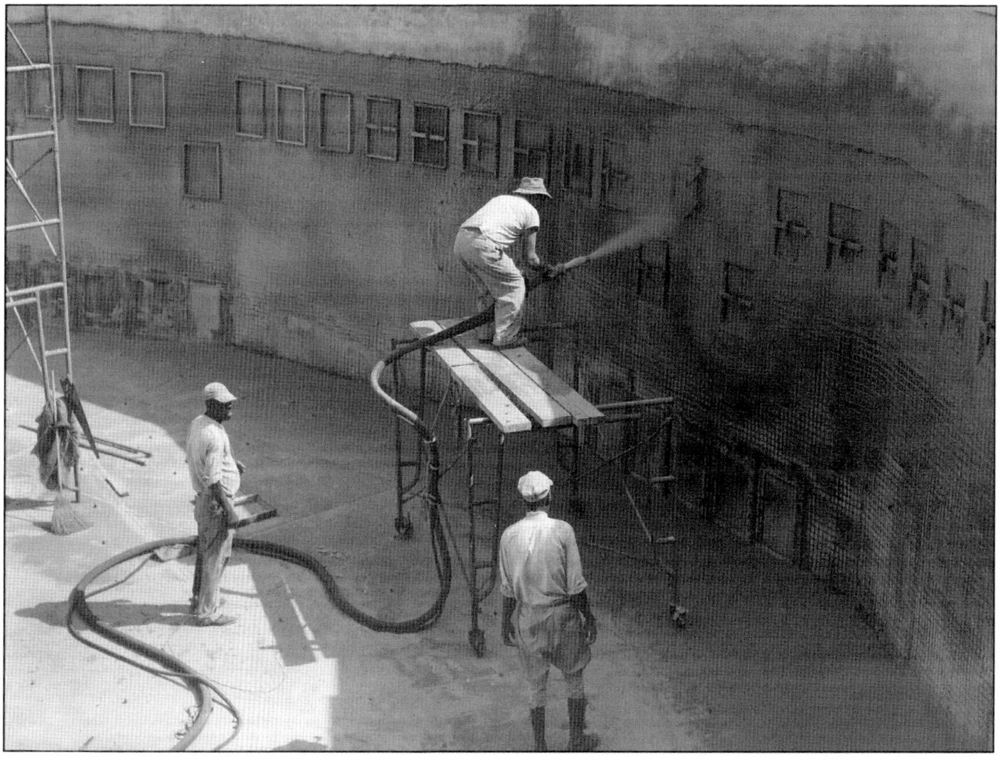

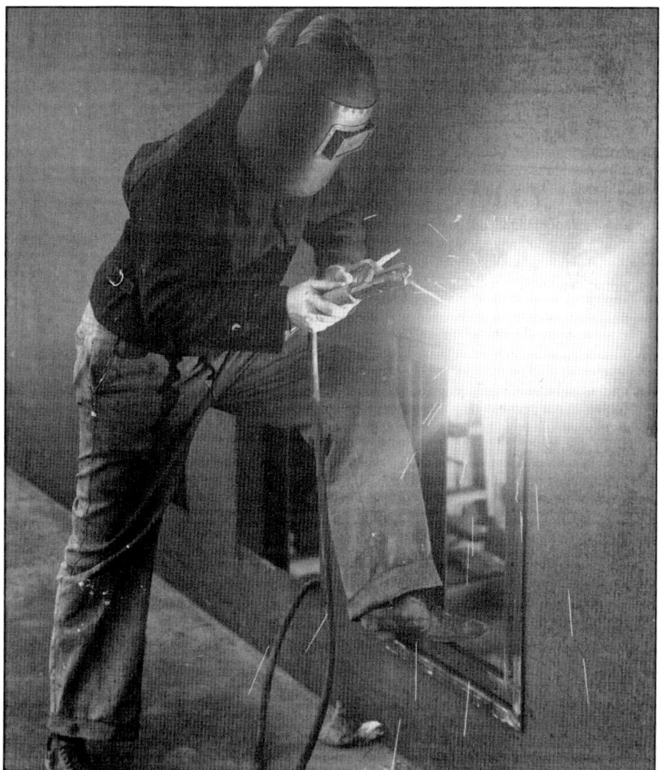

One of the primary challenges for the engineers was to create walls strong enough to support the tremendous weight of the water, estimated to be 3,000 tons in each pool. Contractor Arthur Perry solved this problem by using welded steel plates lined with steel mesh reinforcement from Ingalls Iron Works Company. The absence of rivets delivered the desired results of increasing the strength of the structure and permitting the wall thickness to be reduced, while supplying the added benefit of shortening the building time. The steel was then sprayed with gunite, a mixture of sand and water applied at high pressure, and finished with a newly devised rubber paint sealant in a natural color called "sea green enamel."

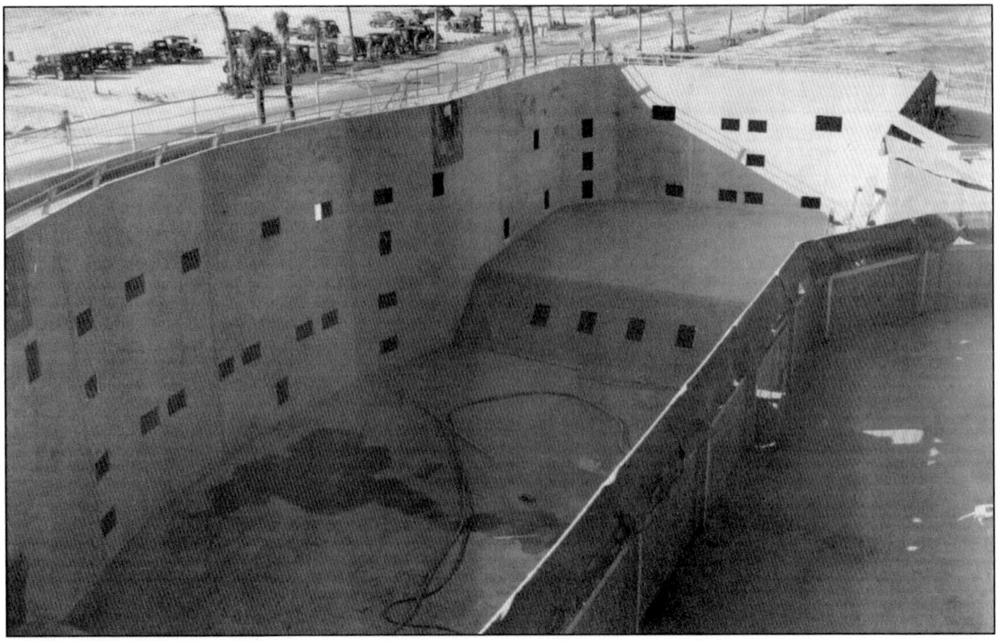

The facility design included two pools, one rectangular and one circular, connected by a flume. The rectangular oceanarium was 100 feet long, 40 feet wide, and 18 feet deep, with a capacity of approximately 450,000 gallons of seawater. This pool would become the primary home of tropical fish and sharks. The circular oceanarium was 75 feet in diameter and 12 feet deep, with a capacity of approximately 400,000 gallons of seawater. This pool would become the primary home of the dolphin and whale family. The vast array of underwater viewing windows were made of specially tempered glass that could withstand pressures of up to 1,800 pounds per square inch. They were selectively placed on three levels so that visitors could choose the best possible view depending on the time of day, the sunlight, and the animals' behavior.

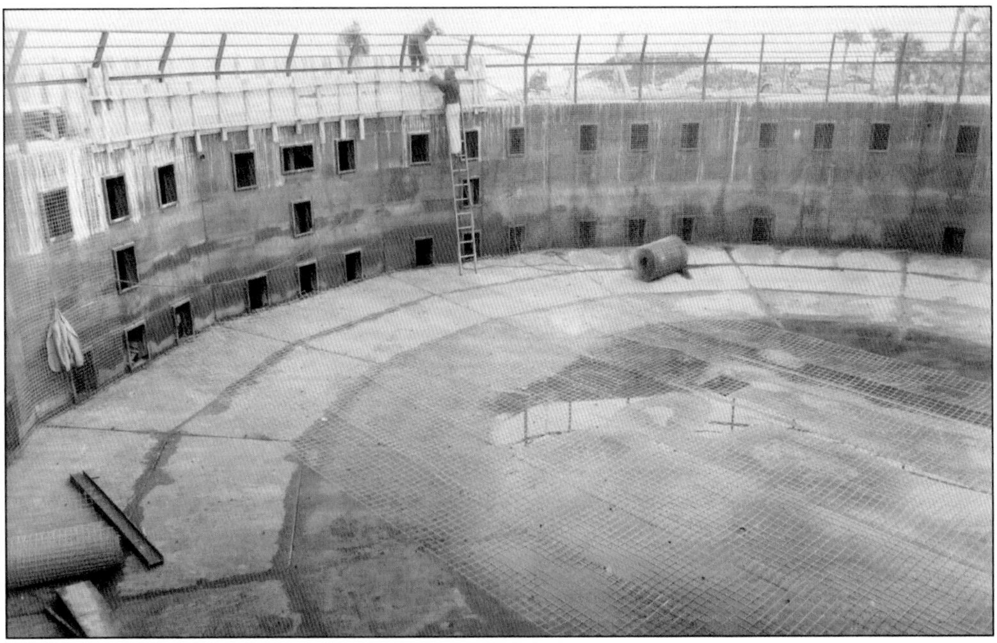

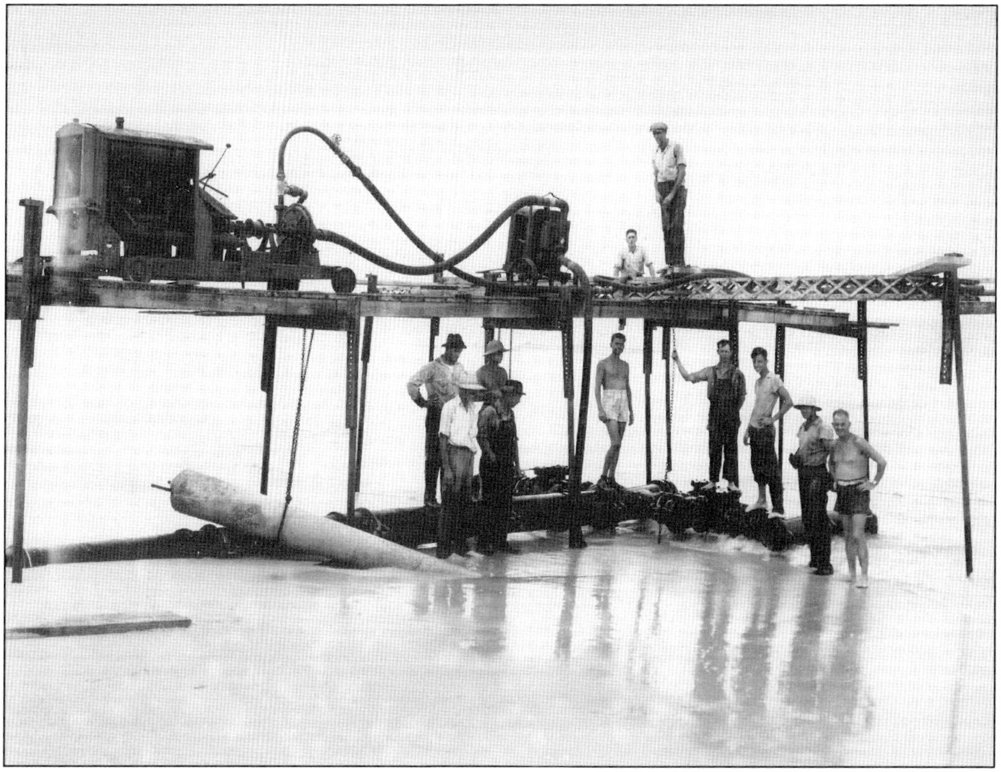

Water clarity in the pools was essential if the facility was to be used as a film studio, and the water quality would be crucial to the health of the animals that lived there. For these reasons, the filtration, exchange, and replacement of seawater in the pools was of vital importance. The water intake pipe was buried under the beach sand and extended out past the surf line, where the seawater would be drawn into the facility by electric pumps through perforated clay pipe, using sand and gravel as natural filters. In the event of a power outage, a gas-powered standby unit kept the system operating seamlessly. Copper compounds were added to the water, improving clarity by controlling algae buildup while killing parasites and diseases common to the tropical fish.

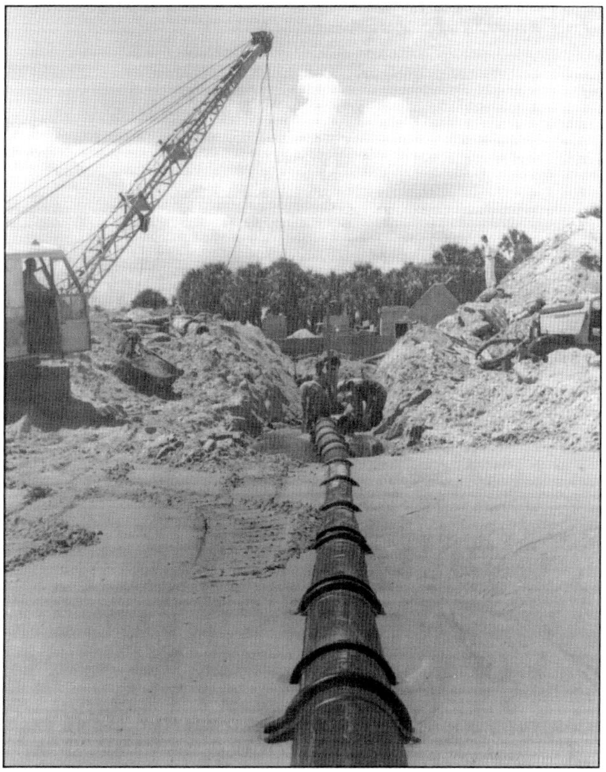

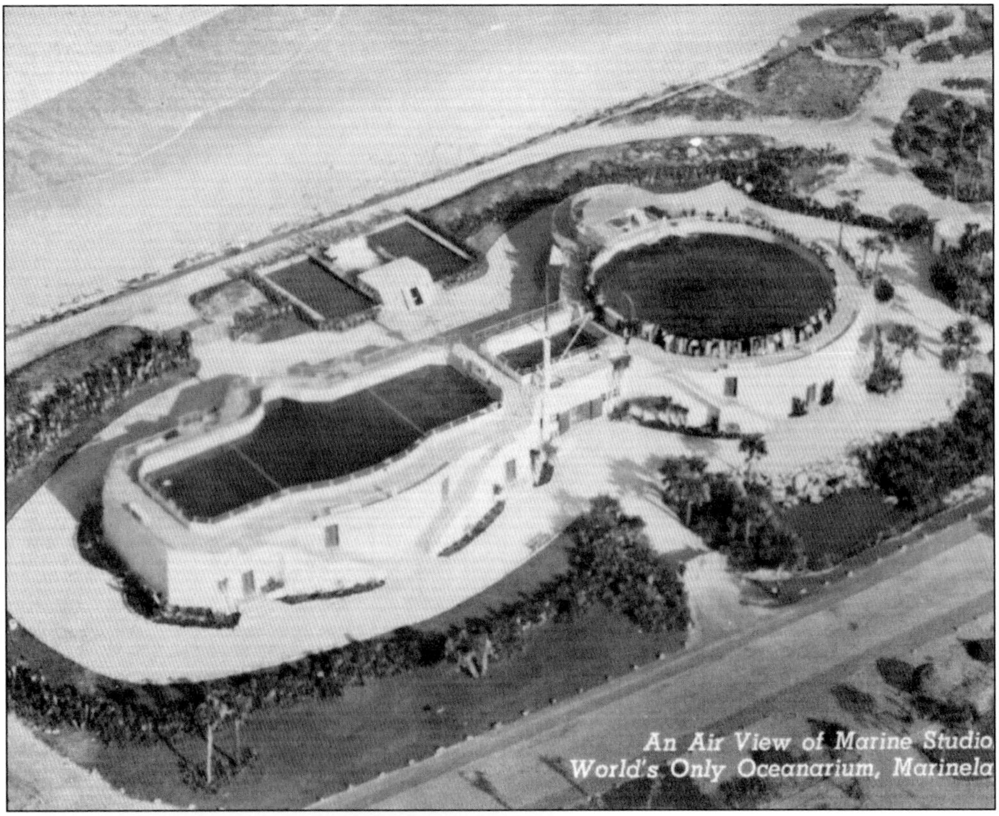

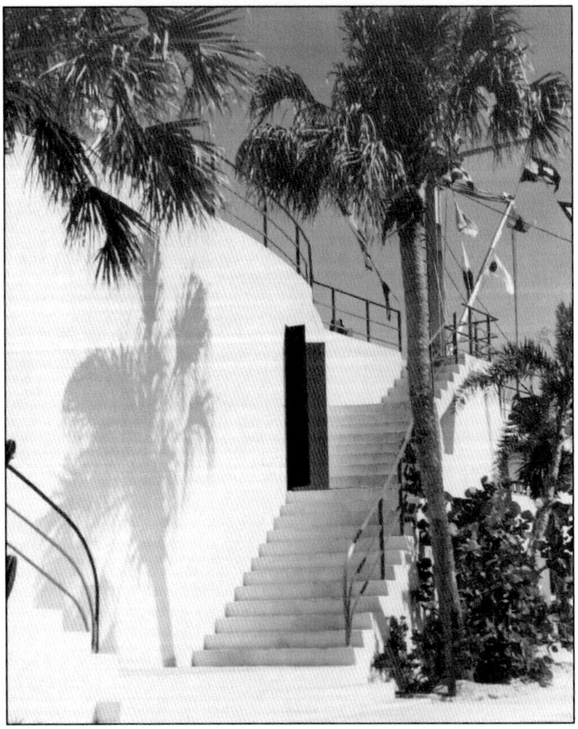

Marine Studios was built directly on Florida's Highway A1A, the only road in the 1930s to span the state from north to south, which conveniently brought potential guests right to the oceanarium's front door. For a fee of $1.10, visitors would be transported into a unique undersea world. The nautical moderne style of architecture was intended to create a shipboard atmosphere, complete with three decks: A deck being an open-air promenade, B deck providing a view halfway down beneath the water, and C deck offering views near the oceanarium floor. Guests were afforded these underwater views of the animals through porthole windows. The addition of nautical flags and sailor-inspired staff uniforms added to the boat-like feel and appearance, all cleverly designed to draw guests into a themed world where the ocean came ashore.

With almost one million gallons of seawater to stock with specimens, Marine Studios' first ocean collection team was led by Capt. Eugene Williams. He captained the oceanarium's ingeniously designed vessel *Porpoise*, equipped with a trapdoor in the stern that led to a large live well, which enabled safe and comfortable transport of large ocean specimens to Marine Studios.

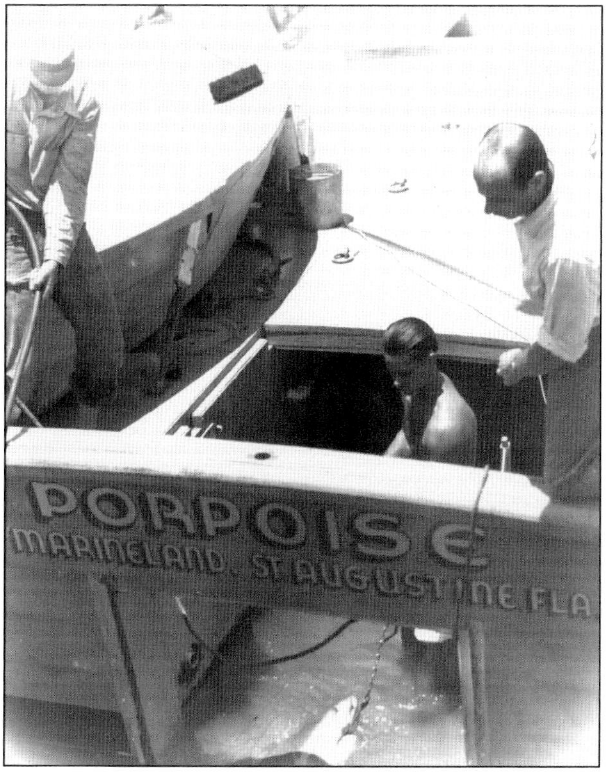

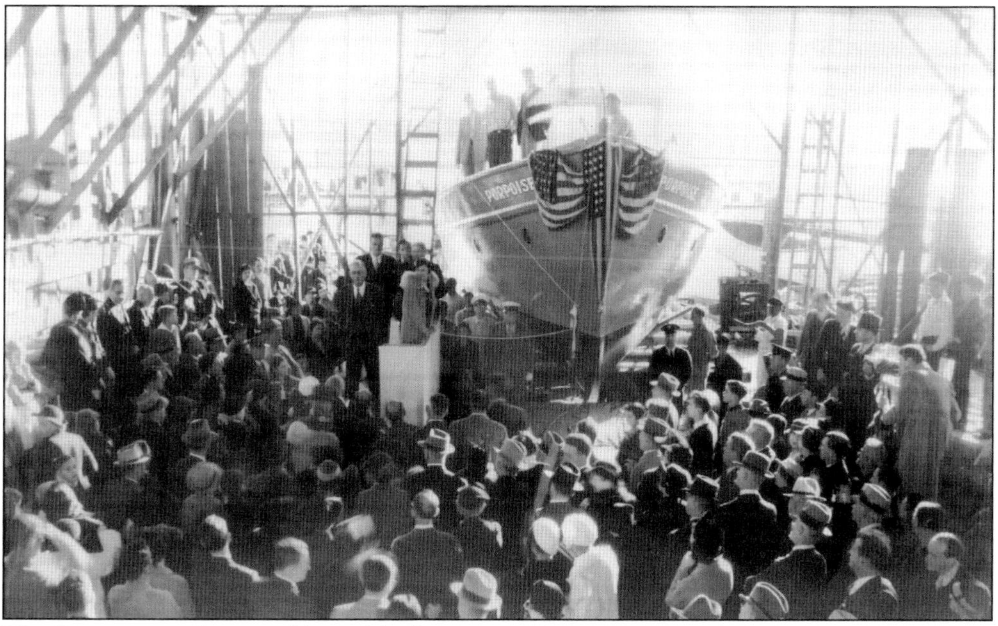

The *Porpoise* was christened by Mildred Victoria Thompson Cone, the Florida governor's wife, in the spring of 1938, ushering in a time of adventure and excitement about the wonders the sea might yield. Local support for Marine Studios flourished, including remarkable donations by resident fisherman Terrel Hayes, who assisted with the original stocking of specimens but refused to take payment for services rendered, content to contribute to this worthwhile endeavor.

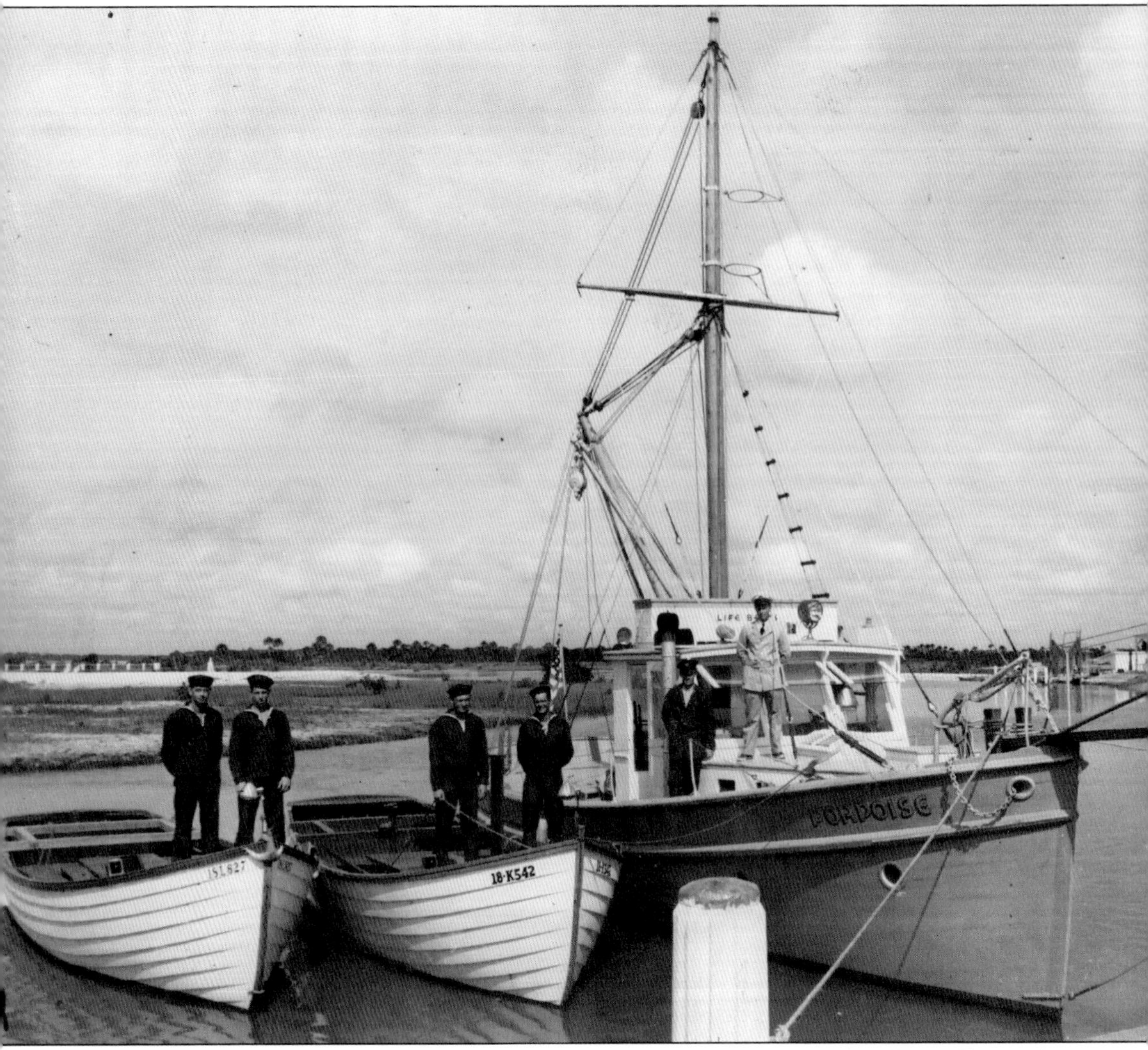

Designed by naval architect E.B. McCrohan, the *Porpoise* was Marine Studios' first collecting boat. This specially designed vessel was adapted from fishing boats commonly used in Cuba and Panama in the 1930s to successfully deliver fresh, living fish to port, an admirable feat in the days before the advent of refrigeration. Captain Williams, whose father served as a guide to famous novelist turned fisherman Zane Grey, safely transported marine animals of all sizes to Marine Studios with the help of the 20-by-4-foot live well found on the vessel. At 48 feet in length, the *Porpoise* was designed to handle up to 2,000 pounds of live specimens at a time. A skiff, the *Seal*, would sometimes accompany the larger ship on collecting trips, shuttling specimens back and forth to the facility so the larger craft could remain at sea. Distinct and separate from the land-based staff, the men who collected specimens were experts in their own right, and their behind-the-scenes efforts launched Marine Studios' success.

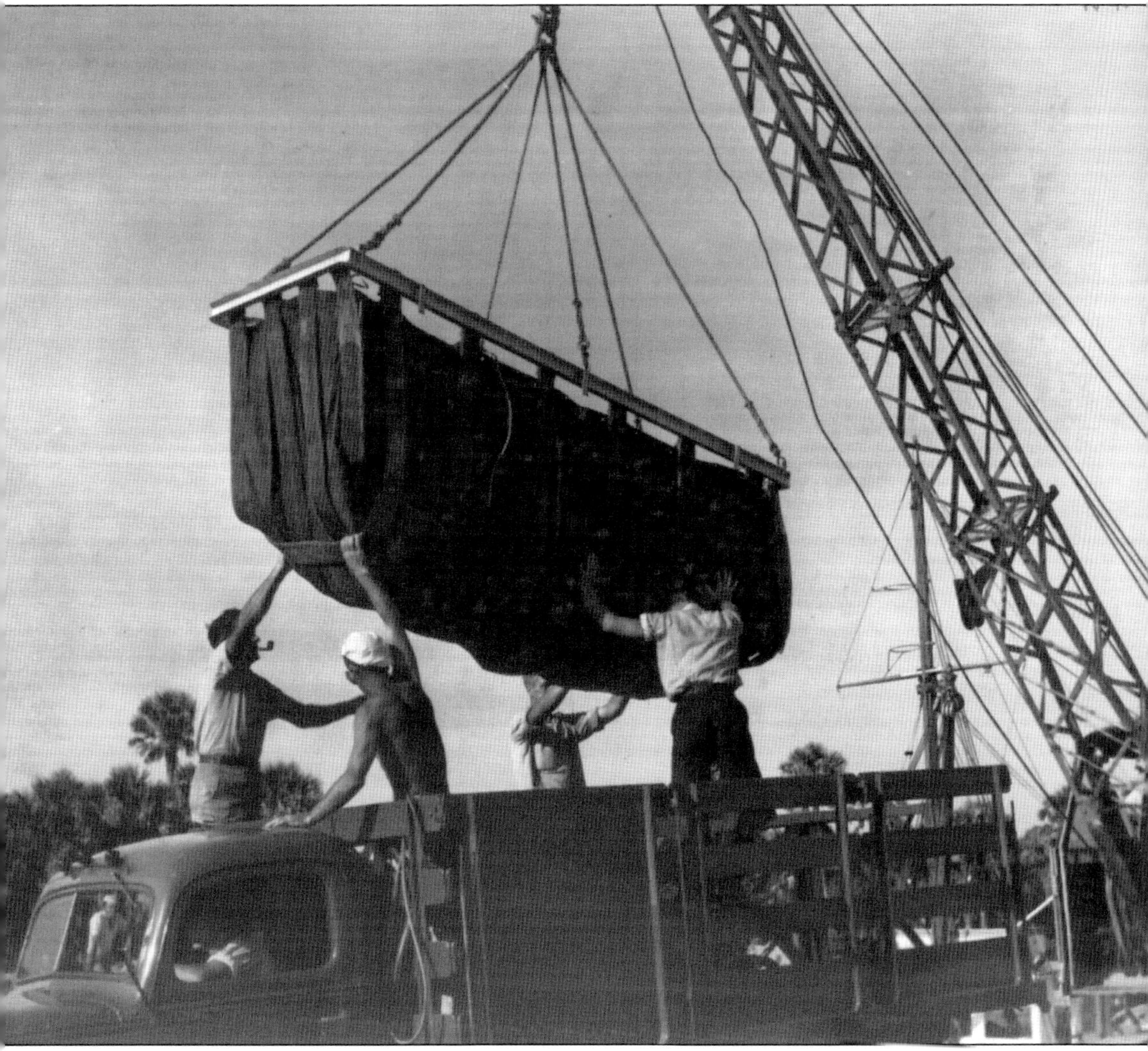

For the initial collection trip in 1938, the *Porpoise* left for an extended three-week fishing expedition to Miami, the Keys, and the Bahama Banks. During the trip, Captain Williams was assisted by Hal P. Davidson, who was in charge of dive operations for the trip as well as the procurement of all shell and coral specimens. This was the first extensive trip on the boat and was used to test all of the specialized and revolutionary equipment on board, her seaworthiness, and the general ability to perform the work she was designed to do. W.H. Gerecke, staff photographer, accompanied the expedition to document the occasion. Specimens were brought to dock every third day and transported by tank truck to Marine Studios. Because the Key West Aquarium was said to be the only aquarium to date to successfully display live tarpon, Marine Studios hired local Key West collectors to assist in obtaining the tarpon for the facility.

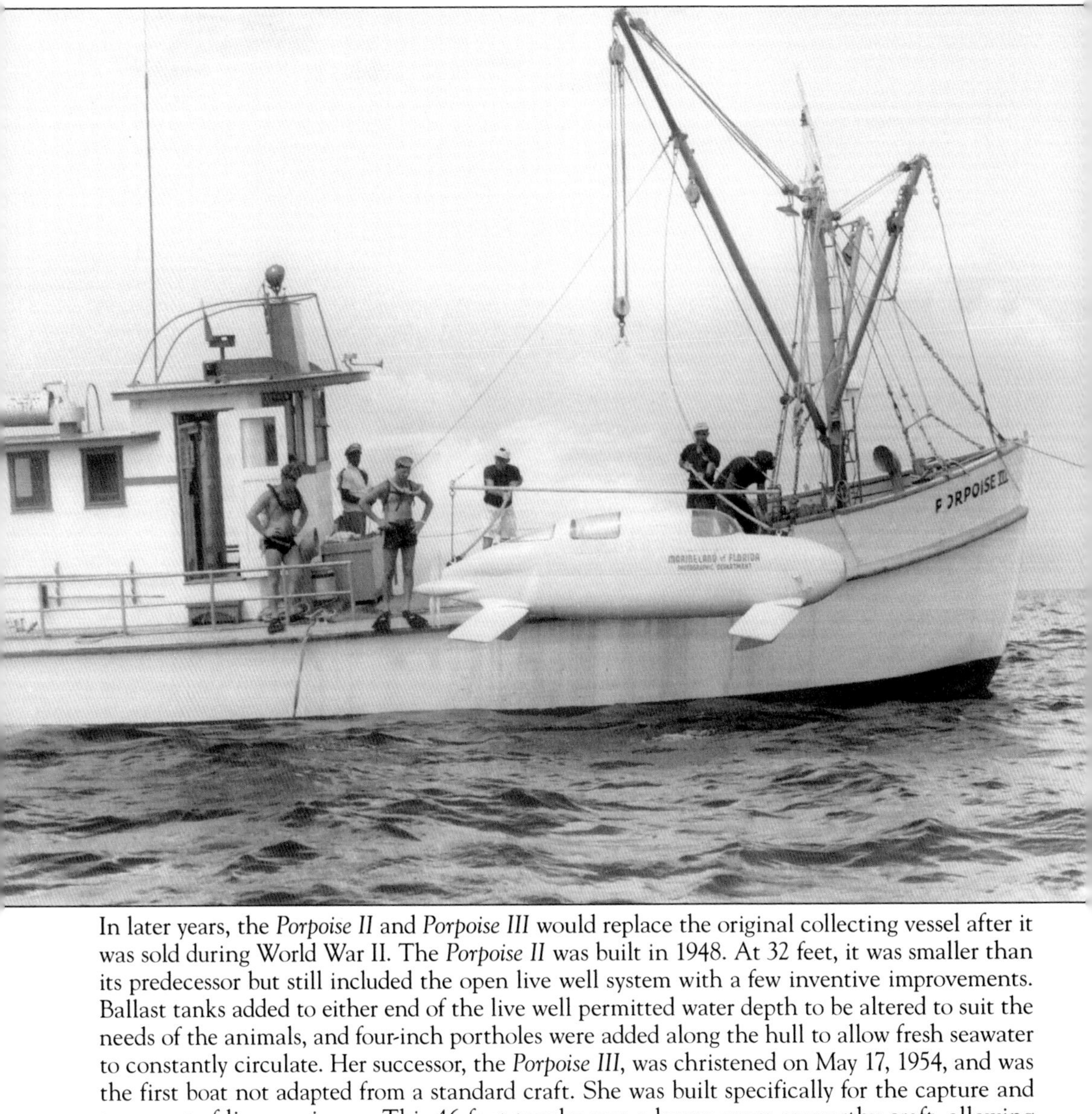

In later years, the *Porpoise II* and *Porpoise III* would replace the original collecting vessel after it was sold during World War II. The *Porpoise II* was built in 1948. At 32 feet, it was smaller than its predecessor but still included the open live well system with a few inventive improvements. Ballast tanks added to either end of the live well permitted water depth to be altered to suit the needs of the animals, and four-inch portholes were added along the hull to allow fresh seawater to constantly circulate. Her successor, the *Porpoise III*, was christened on May 17, 1954, and was the first boat not adapted from a standard craft. She was built specifically for the capture and transport of live specimens. This 46-foot trawler was a larger, more seaworthy craft, allowing collecting operations to be extended out to the Gulf Stream while providing additional comfort to Capt. Ronnie Capo, including complete facilities for a two-man crew, a ship-to-shore radio, an autopilot setting, and dual controls on the flying bridge.

The ability to collect rare specimens from the ocean was only the beginning of the adventure. Whether large or small, the complexities and equipment necessary to transport, acclimate, and care for strange and exotic ocean creatures presented even bigger challenges. Here, Marine Studios' general manager Bill Rolleston assists in securing a newly collected animal with the use of a specially designed transport sling.

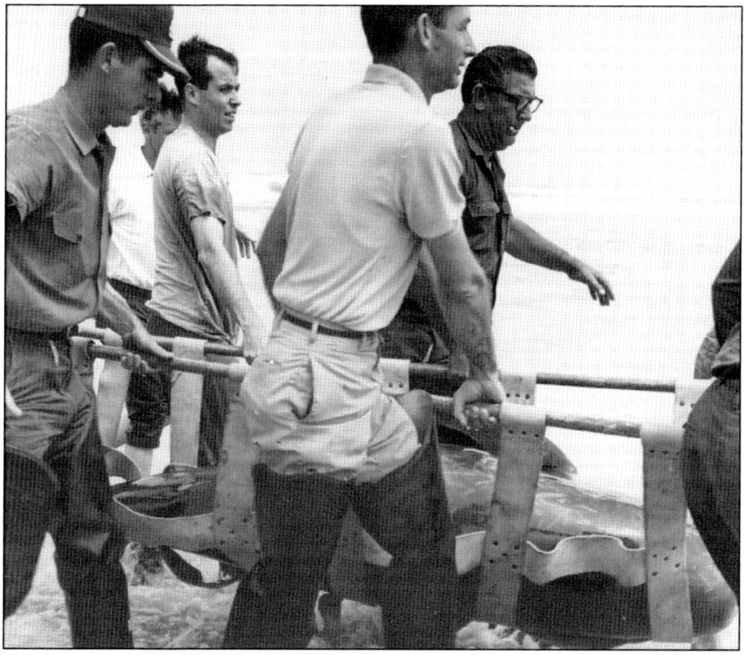

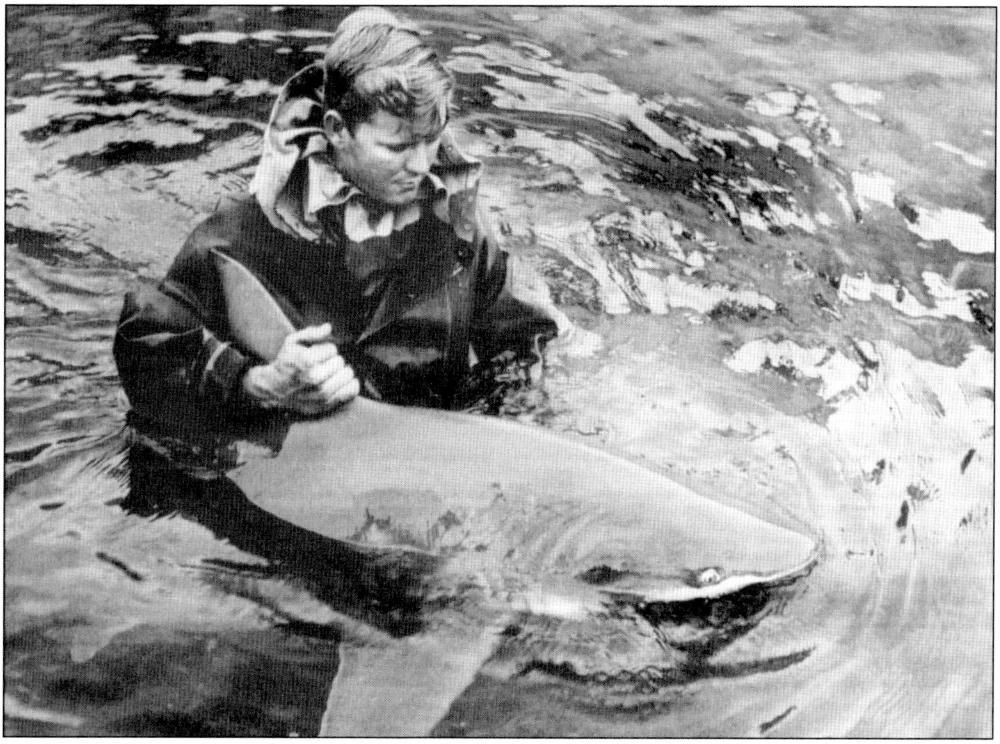

To improve the welfare of sharks and the safety of staff during transport, Marine Studios scientists, in partnership with the American Museum of Natural History, perfected a method of shark anesthesia. An innovative technique for its time, it required the development of a hypodermic injection system and the identification of effective medications that lacked negative side effects. Divers were often required to "walk" the sharks until the anesthesia wore off.

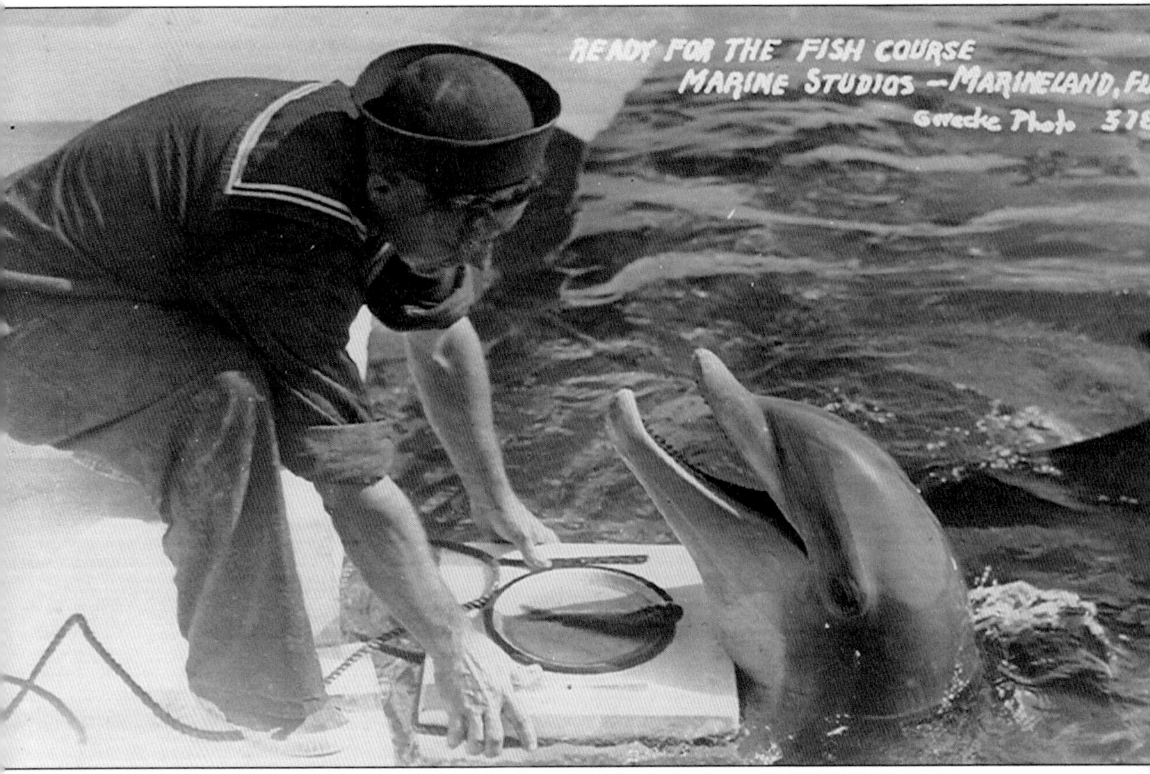

Originally, the Marine Studios collection fleet was not only responsible for stocking the oceanarium, but also for bringing in fresh food to feed its inhabitants, usually in the form of blue runner, butterfish, mullet, and sardines. This was a daily practice that required an enormous amount of staff time and labor to accomplish. In 1946, less than 10 years later, Marine Studios upgraded its food preparation areas by installing state-of-the-art refrigeration and freezer facilities on site that were large enough to store a full year's worth of frozen menhaden, blue runner, and squid, which totaled approximately 85,000 pounds of food. Once fully operational, the fleet's only food responsibility was to continue to provide fresh mullet for the jumping dolphins. Although approximately 90 percent of Marine Studios' food fish was now purchased frozen from commercial dealers, there was a long-held misconception by both private and commercial fishermen in the area that Marine Studios was personally responsible for the decline in their catches.

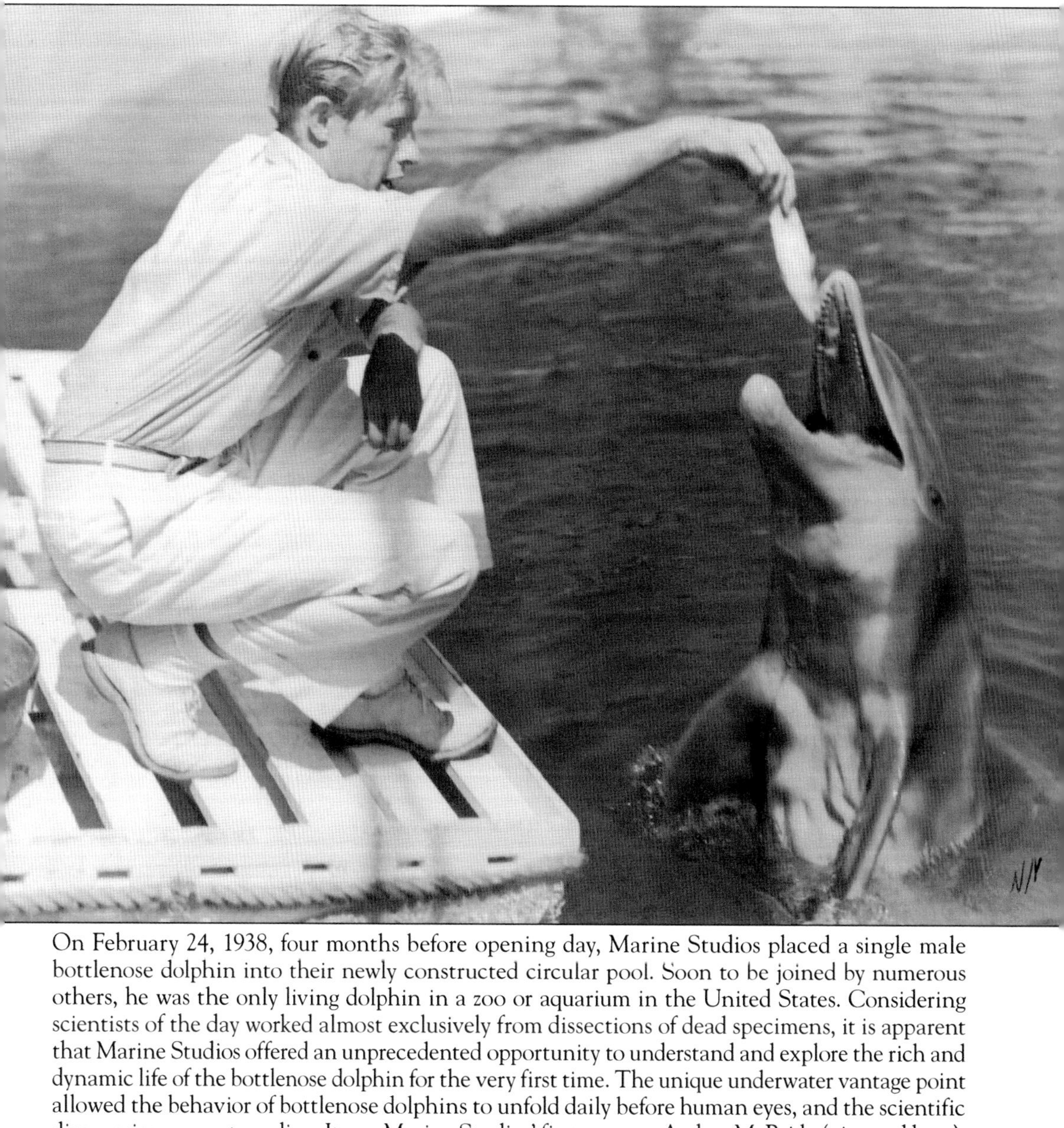

On February 24, 1938, four months before opening day, Marine Studios placed a single male bottlenose dolphin into their newly constructed circular pool. Soon to be joined by numerous others, he was the only living dolphin in a zoo or aquarium in the United States. Considering scientists of the day worked almost exclusively from dissections of dead specimens, it is apparent that Marine Studios offered an unprecedented opportunity to understand and explore the rich and dynamic life of the bottlenose dolphin for the very first time. The unique underwater vantage point allowed the behavior of bottlenose dolphins to unfold daily before human eyes, and the scientific discoveries were astounding. It was Marine Studios' first curator, Arthur McBride (pictured here), who has been credited with discovering much of what is now known, and subsequently taken for granted, about bottlenose dolphins, including everything from basic biological information about how they breathe, sleep, and mate to more complex behavioral observations concerning dolphin social orders, hierarchy, and dominance displays.

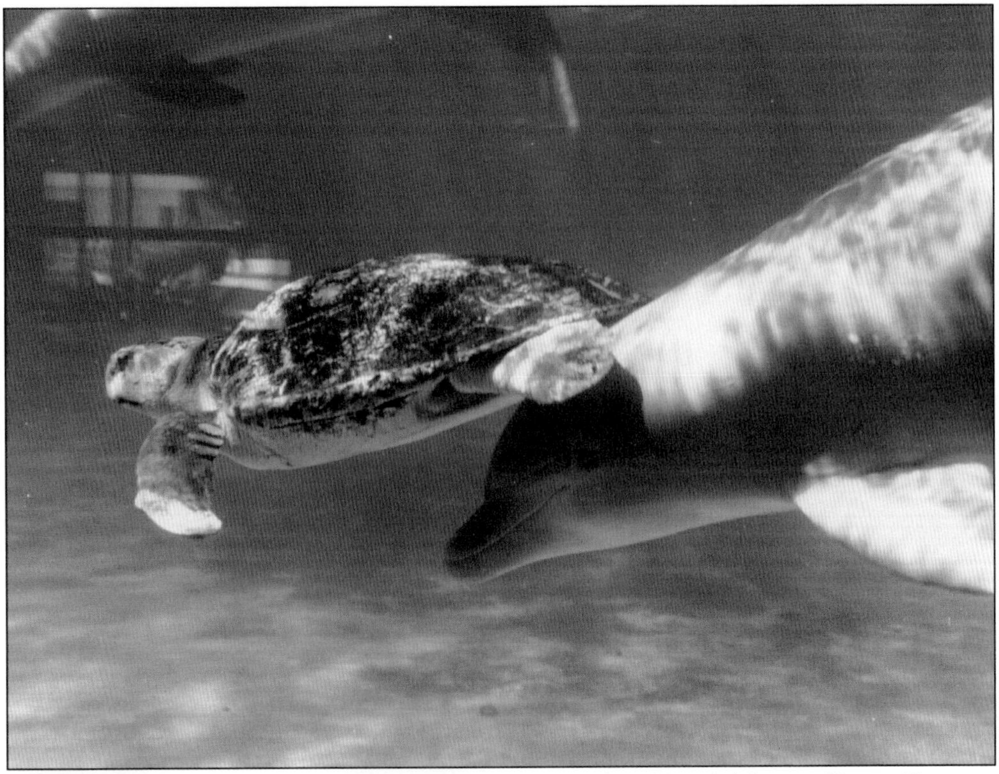

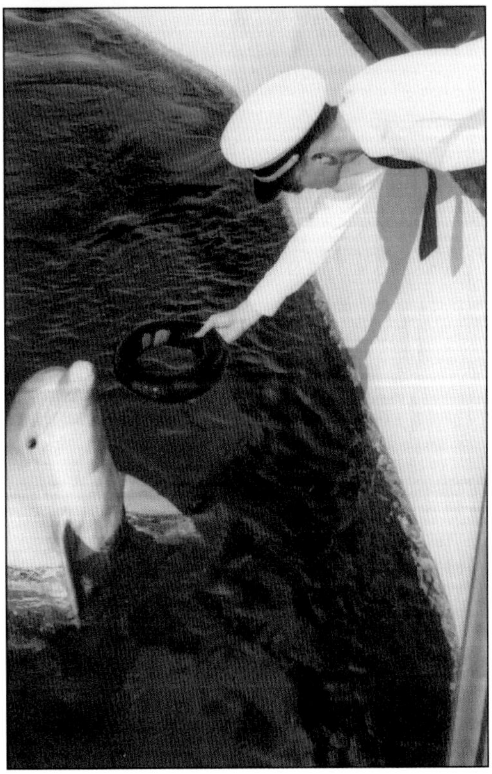

In March 1938, a dolphin and her calf were introduced to the circular oceanarium, and the calf immediately initiated play. Her favorite game was pushing a sea turtle around the pool. Although enriching, the sea turtles needed some relief, so the staff introduced the very first man-made toys in the form of rubber inner tubes and balls. The calf invented many of her own games and frequently initiated play with the staff, soliciting body rubs and playfully nibbling at the divers that entered her pool. Not surprisingly, she was the first dolphin to take fish underwater directly from a diver's hand and quickly became everyone's favorite animal. Referred to as their pet, she was the first Marine Studios inhabitant to be personalized with a name, affectionately called "Baby."

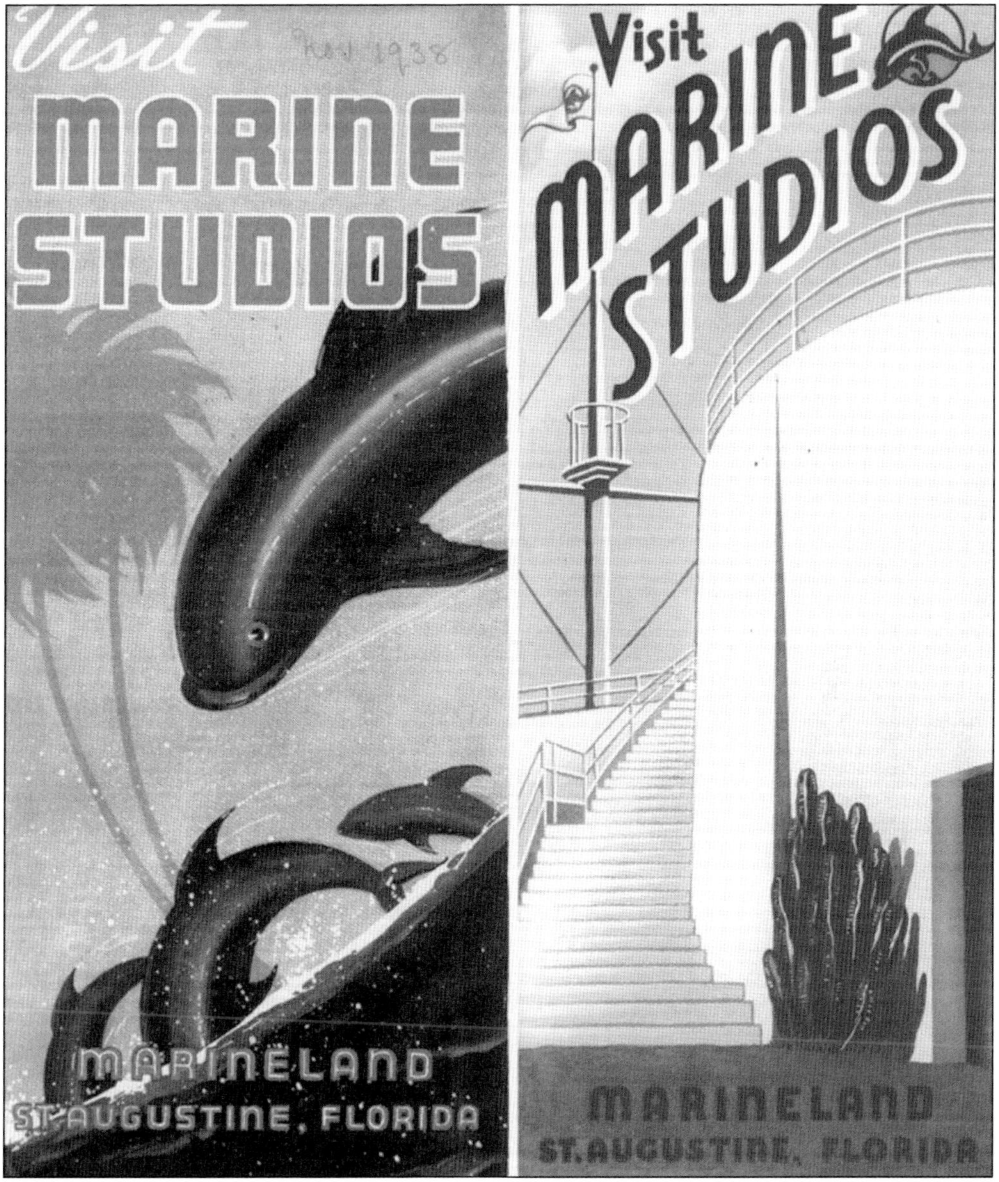

In April 1938, the facility was partially opened to the press for an inside peek into one pool that contained dolphins and a sawfish. This early opening for the press resulted in rave reviews and priceless free publicity. Marketing and promotional materials, such as these brochures, were also developed, printed, and distributed to entice crowds to visit the new attraction. With a complete lack of photographs, they were filled with artist renderings and depictions of what could be experienced at the facility. By June, Marine Studios was ready to open to the world. With construction complete and the pools stocked with animals, the grand opening was imminent. The founding group as well as the large staff—many of whom had relocated great distances for a chance to work on this groundbreaking project—were unsure of what to expect. The original investors, including Whitney, Burden, and Pratt, were eager to see if their venture had been a sound one.

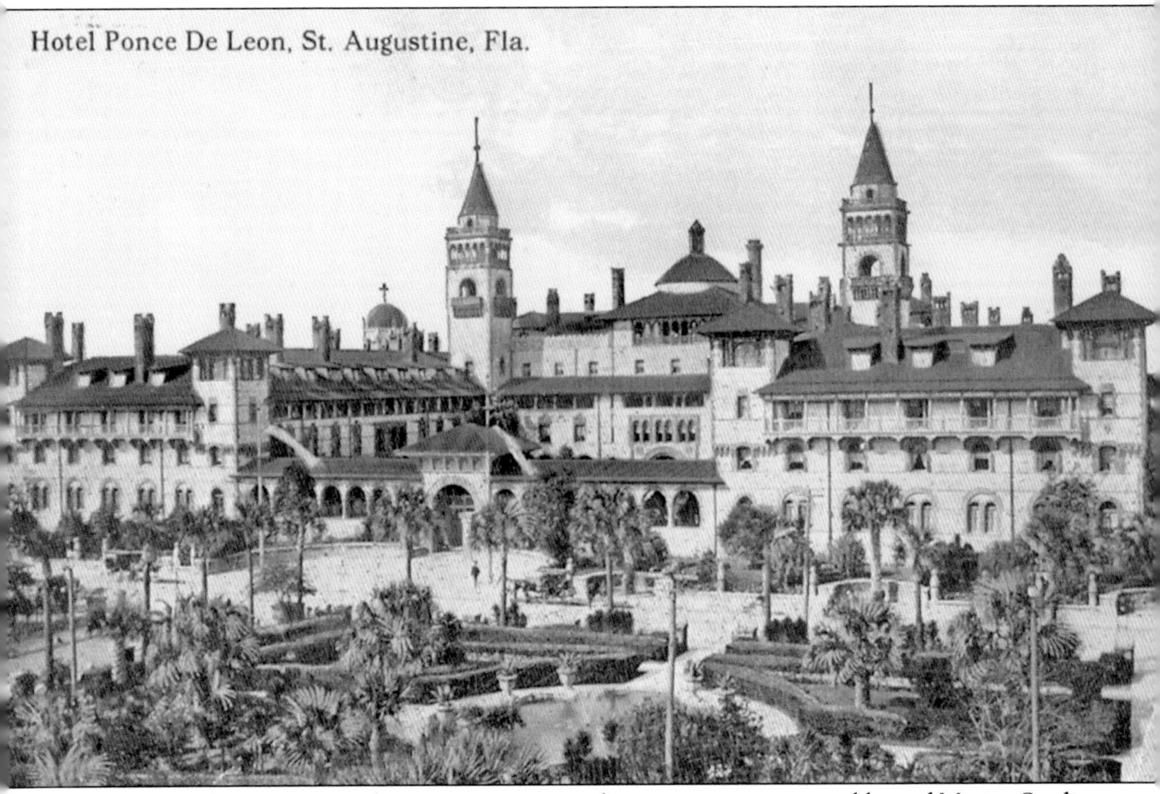

With much promotion and press, excitement in the community was tangible, and Marine Studios officials were honored in March 1938 at an elaborate dinner at the Hotel Ponce de Leon (today part of Flagler College) to celebrate the anticipated opening of the facility. Billy Marshall led the Meyers Davis Orchestra, dressed in crew costumes, from a stage transformed into the bow of a boat, entertaining the crowd with nautical-themed songs. Additional entertainment was provided by the Skilling Sextette (a family singing group) and the Ponce de Leon Strollers. Guests entered the room along a gangplank to tables adorned with seashell centerpieces and nautical paper hats and dined on pompano, lobster, green turtle soup, ice cream, and cakes shaped to look like fish. The most outstanding achievement of the evening was an ice sculpture of Marine Studios, created by headwaiter Adolph Bittner, which was surrounded by a buffet table of fish in aspic. All guests received fish and penguin party favors to take home as a souvenir of the evening.

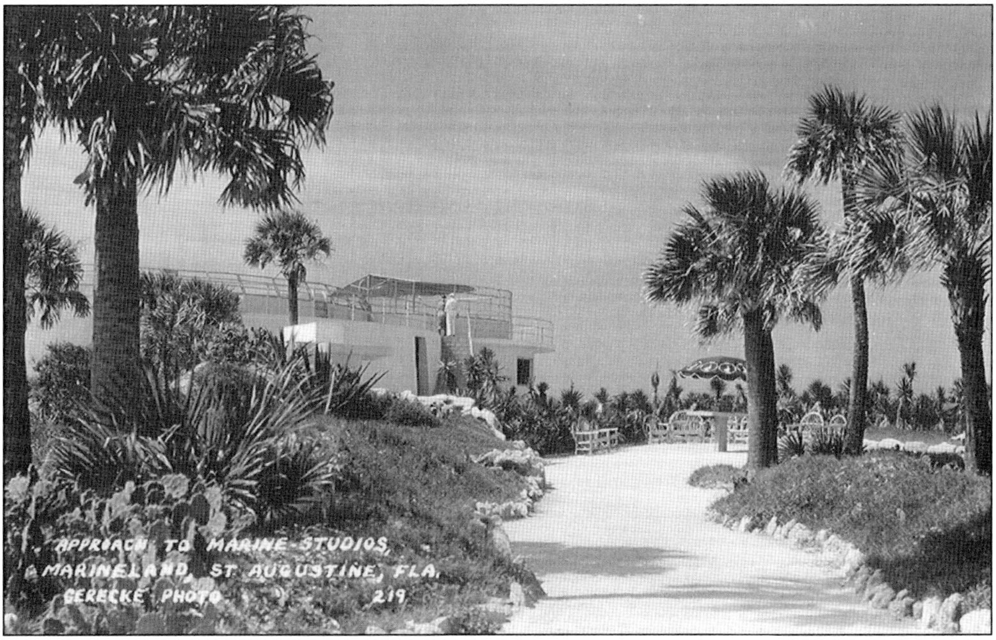

With the animals in place and the main portion of the facility complete, it was time to focus on the finishing details. For this, landscape architect Mulford B. Foster was hired, and under his direction, the lush tropical theme of Marine Studios was brought to life. The main entrance was transformed into a palm-filled park that offered a pleasant atmosphere for guests to spend time strolling along sand-covered walkways lined with coquina rock borders and enjoy blooming flower gardens, refreshing drinking fountains, shaded benches, and breathtaking views of the beach and ocean. The entrance gate, marked by a great rock tablet engraved with the Marine Studios logo, had two flags flying above it, officially welcoming visitors to the world's first and only oceanarium.

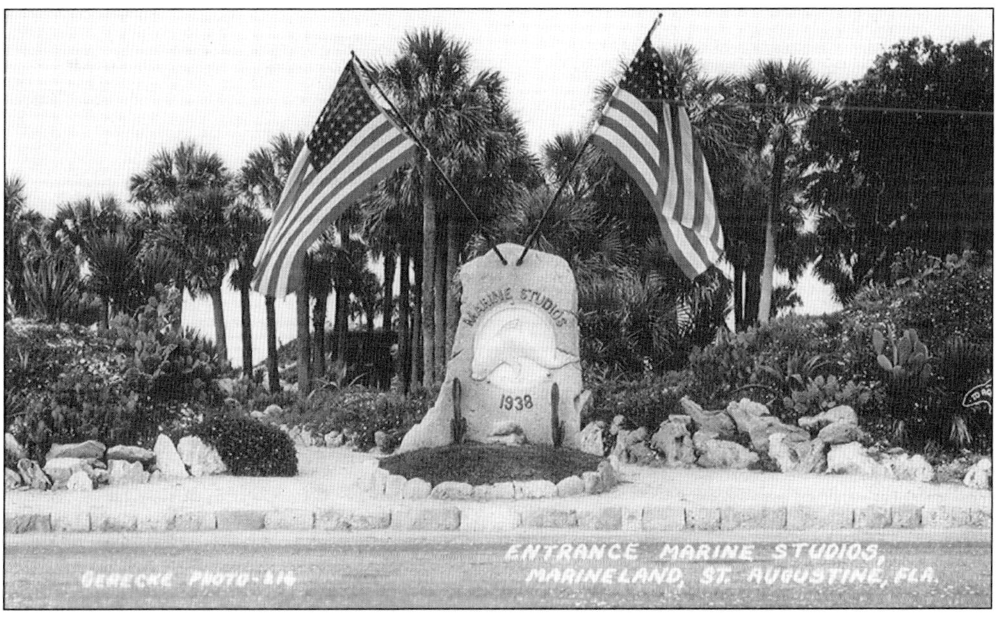

June 24, 1938

Nearly 30,000 People Jam Ocean Front As Marine Studios Open

A Portion of the Tremendous Crowd

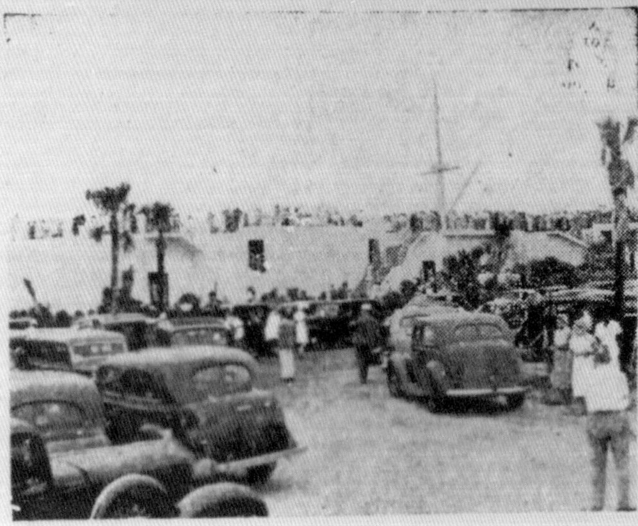

Caught by the camera in the above picture is a portion of the record-breaking throngs which massed from many localities yesterday at Marineland to witness the formal dedication of the $500,000 Marine Studios. This scene, which gives an indication of the size of the crowd, estimated at nearly 30,000, was snapped from the western side of the Ocean Shore Boulevard, looking eastward across the automobile-packed highway to the gigantic tanks swarming with men, women and

Traffic Blocked for Miles By Huge Throng Gathed for Ceremony

NOTABLES SPEAK

Spectacular Tourist Attraction Is Thrown Open to Public

The pioneer development inaugurated many years ago by Henry Morrison Flagler, creator of Florida's famed East Coast, might be said to have reached culmination yesterday—through the disclosure to a fascinated, thrilled public of the modern-day miracle that is Marine Studios.

Despite the torrential downpour just prior to the opening ceremonies crowds conservatively estimated at between 25,000 and 30,000 people blocked traffic for miles along the Ocean Shore Boulevard to the north and south of Marineland. Massed cars stood for hours unable to move an inch, yet their occupants apparently did not think of departing before getting at least a glimpse of the undersea wonders revealed by the world's largest and

On June 23, 1938, Marine Studios opened to the public. Reservations and apprehension concerning the attendance numbers for this first day dissipated as an estimated 30,000 visitors tried to make their way down the narrow A1A highway. Traffic was literally backed up for miles in both directions, forcing many curious and disappointed ocean enthusiasts to turn away, physically unable to reach the facility. With no precedent for an underwater studio such as this, the public appeal was unexpected, and the overwhelming success of the oceanarium came as a complete surprise. The founders basked in their achievements as Mrs. Douglas Burden led the official ribbon-cutting ceremony. Presiding at the opening ceremony were two U.S. senators, the mayors of St. Augustine and Daytona Beach, the president of the University of Florida, and the president of the Florida Chamber of Commerce. On this seemingly ordinary summer day in Florida, people were just beginning to realize the enormity of what had been accomplished by Marine Studios.

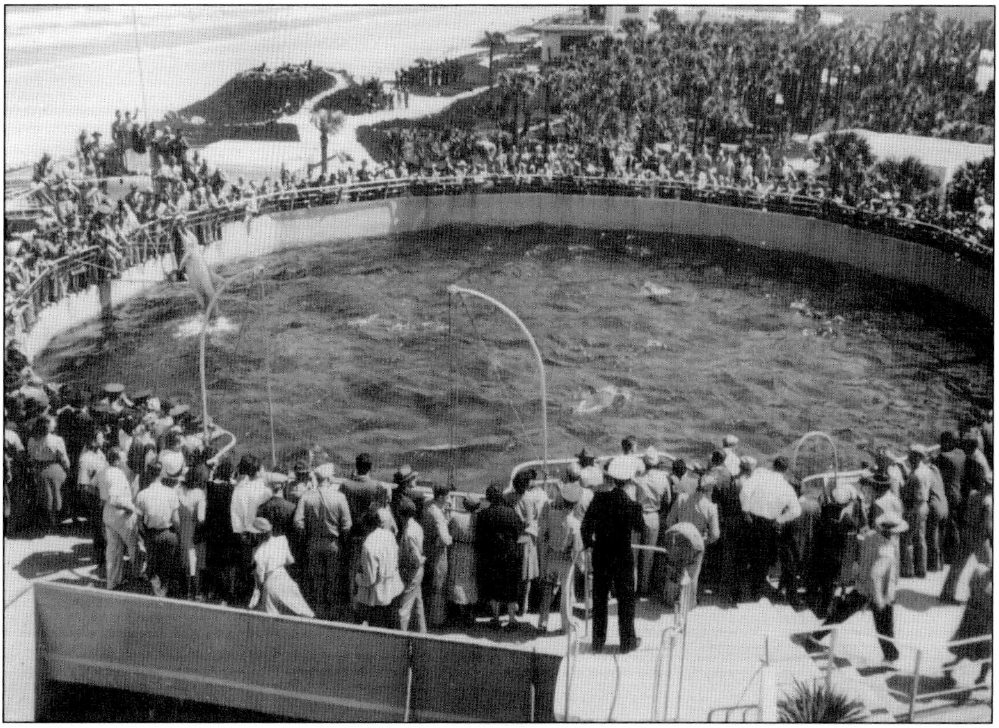

The press swarmed around Marine Studios, and newspapers all across the country carried the exciting news of the marvels it contained. The local *St. Augustine Record* outdid them all, featuring a 42-page spread dedicated solely to the opening celebration. However, it was the *Florida Times Union* that boasted the loudest, informing its readers that they considered Marine Studios to be the ninth wonder of the world and insisting that a visit there would make one doubt the importance of the other eight. At opening day, Marine Studios was already being portrayed as an outstanding biological research center, one of the most popular tourist attractions in the world, and a facility with sound scientific appeal and educational value. This combination of science, entertainment, and educational value is the standard by which zoos and aquariums base their success even today.

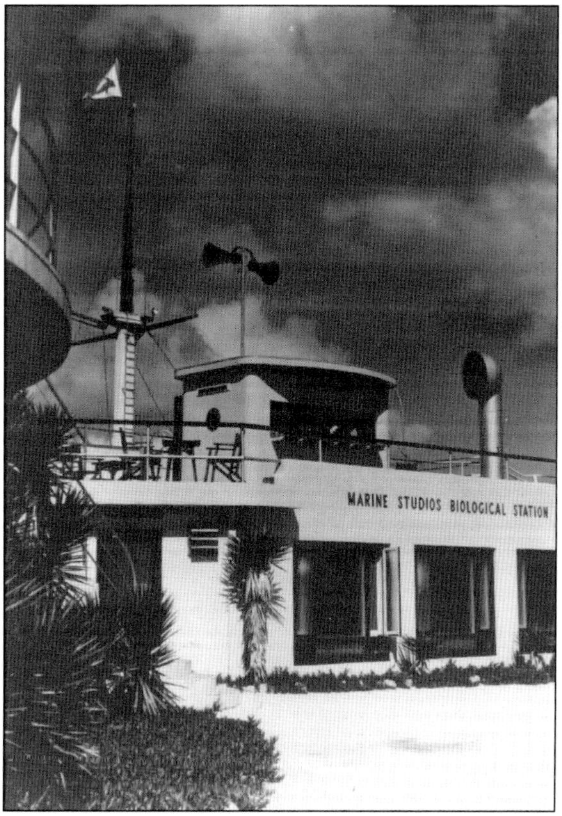

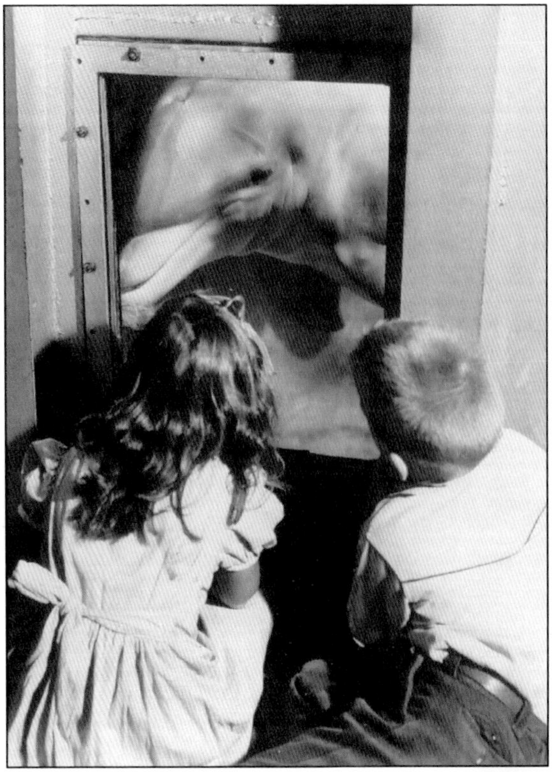

Children and adults alike delighted in the opportunity to watch animals in the oceanarium through the porthole windows. Primarily designed for film and photography, the portholes worked just as well as windows to the sea for an otherwise land-bound public. Here, two young children come face to face for the first time with an Atlantic bottlenose dolphin.

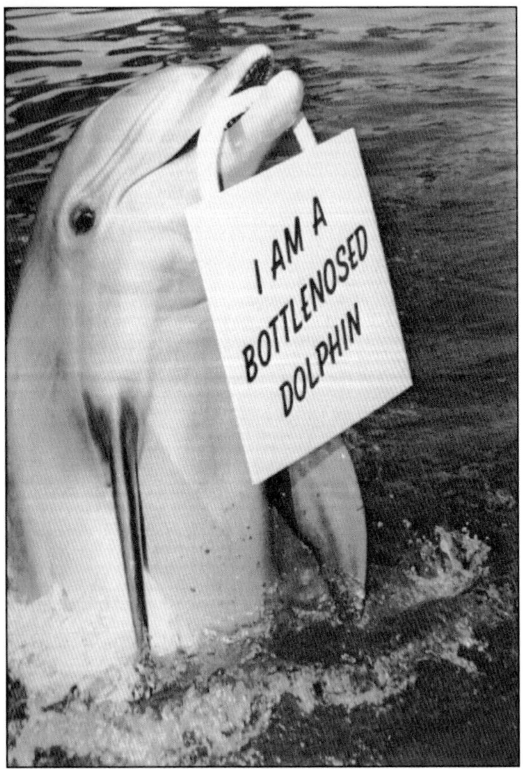

From the top deck, visitors enjoyed a clear and unobstructed view of the dolphins that were hand-fed daily. Although the dolphins were referred to as porpoises in the early years, they were actually bottlenose dolphins. While Marine Studios never displayed porpoises, the staff used the term "porpoise" to differentiate the mammals from the dolphin fish, also known as mahimahi or dorado.

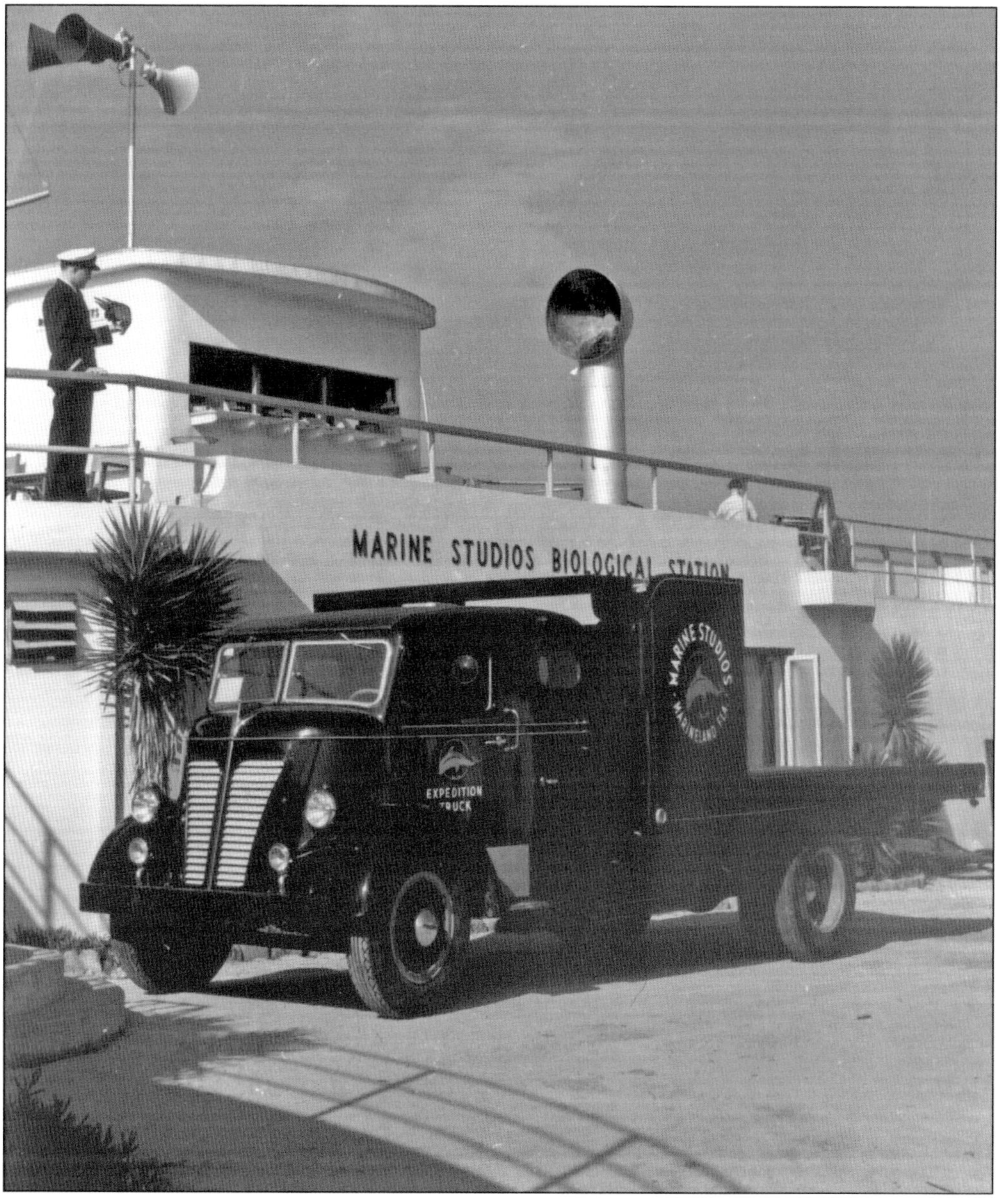

While the dolphins seemed to naturally steal everyone's attention, other Marine Studios species made the headlines from time to time, including a particularly comical hyacinth macaw named Indigo, held here by one of the staff members. Indigo was sent as a gift from C.V. Whitney, who obtained the parrot from the New York Zoological Society. The parrot was provided an open thatch-covered perching area. With complete freedom, Indigo could usually be seen flying up and down the beach with a band of pelicans during the day, typically returning to his perch near nightfall. One evening, Indigo did not return. The worried staff searched for him until a newspaper article featuring Indigo explained his disappearance. The macaw had left his shelter for the shoulder of Fred Goodwin, who was driving on Highway A1A in his convertible, whistling birdcalls he had learned during travels in South America. Indigo, apparently attracted by the calls, remained on Goodwin's shoulder for the entire car ride home. The newspaper article featuring Goodwin's amazing parrot story triggered the return of Indigo to Marine Studios.

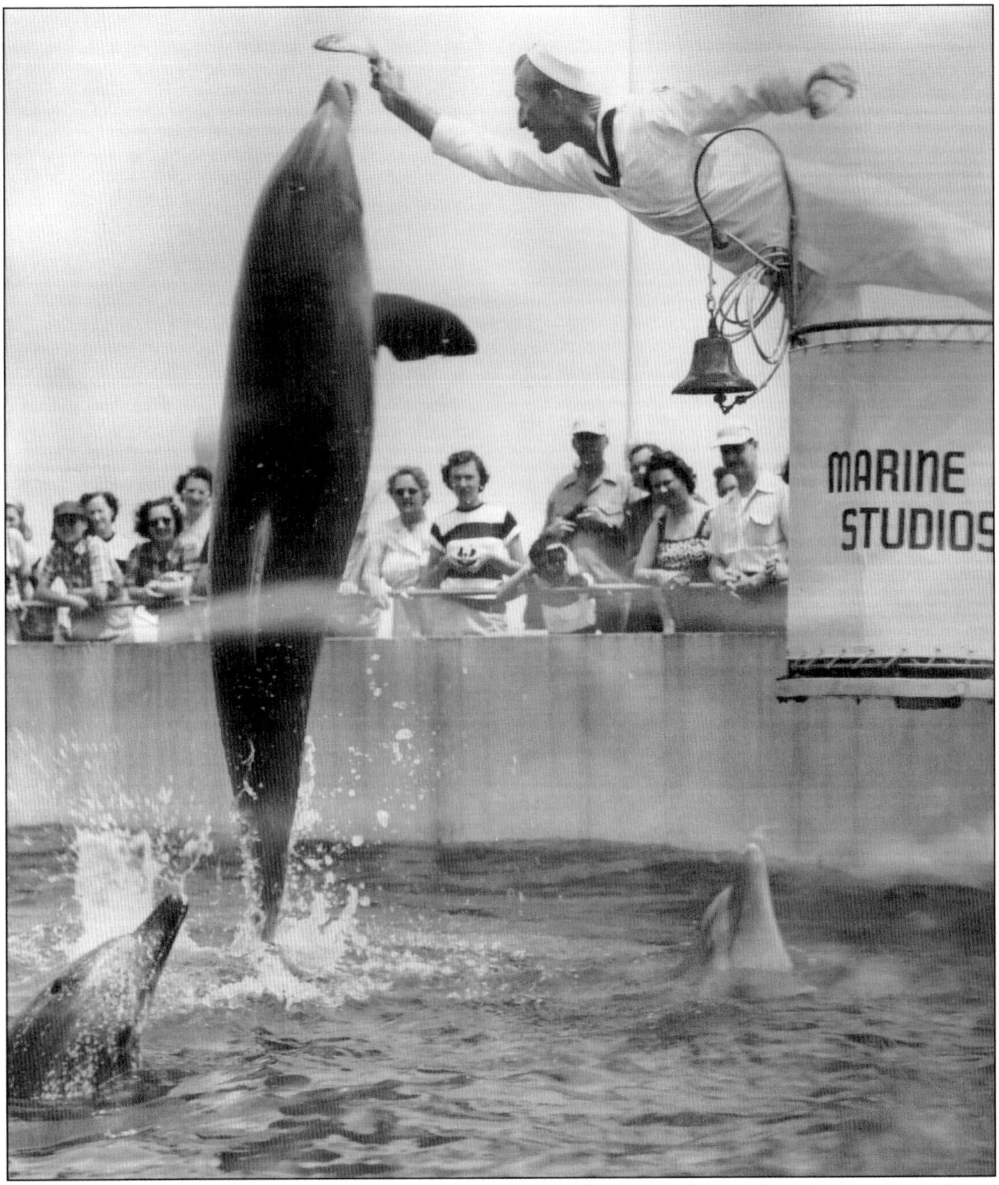

Throughout the wildly successful opening, Marine Studios never lost sight of its purpose. From its inception, the oceanarium held strong to a policy of presenting entertainment without detracting from the educational value of its programs and exhibits. Guides and announcers provided scientifically accurate information about specimens in the oceanariums to the public, while personnel in all areas were prepared to intelligently answer any potential questions. Divers, scientists, businesspeople, and ticket takers were all trained to educate the visiting public about the facility and its animals. Many guests were interested not only in the never-before-seen animals, but also in the career paths of the people who worked with them, especially the divers. Even the technology behind the physical plant operation, bringing water into the facility, and the husbandry necessary to maintain the animals' health was as mysterious to the guests as the animals themselves.

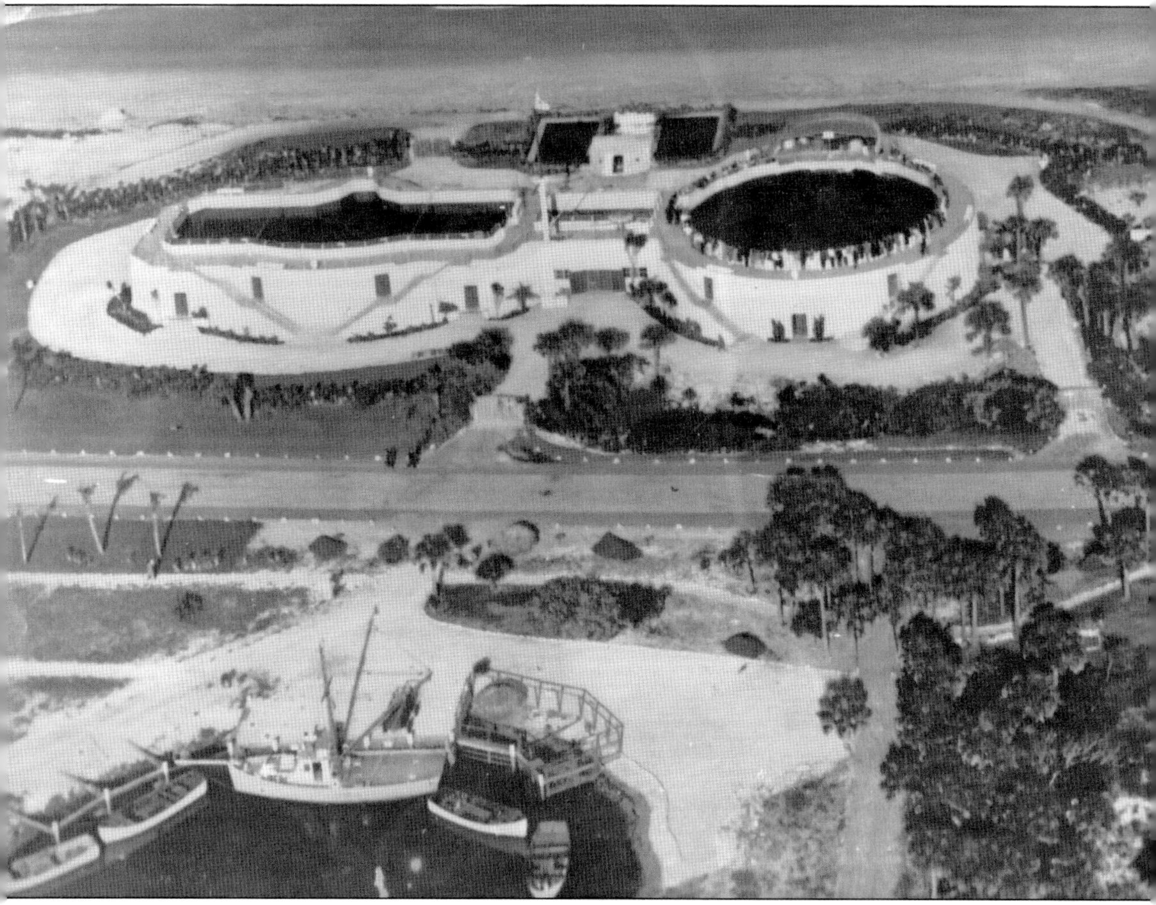

Beyond just providing an entertaining and educational venue to visit, Marine Studios was responsible for many positive contributions and changes to the surrounding community. With the need for phone and power lines along the highway to support the oceanarium, cottages in the neighborhood also received telephone service and electricity for the first time. While these impacts were coincidental, Marine Studios provided intentional benefits to the community through locals-only discounts to the oceanarium. In a true act of self-preservation, representatives of the oceanarium formed a group called the Florida Free Highways Association, which viewed the construction of a proposed toll road between Jacksonville and Miami as destructive to property values along the state's east coast. Instead, they argued that US Route 1 should be turned into a four-lane road and that Highway A1A should be improved. Marine Studios' director of public relations, John Dillin, led the charge.

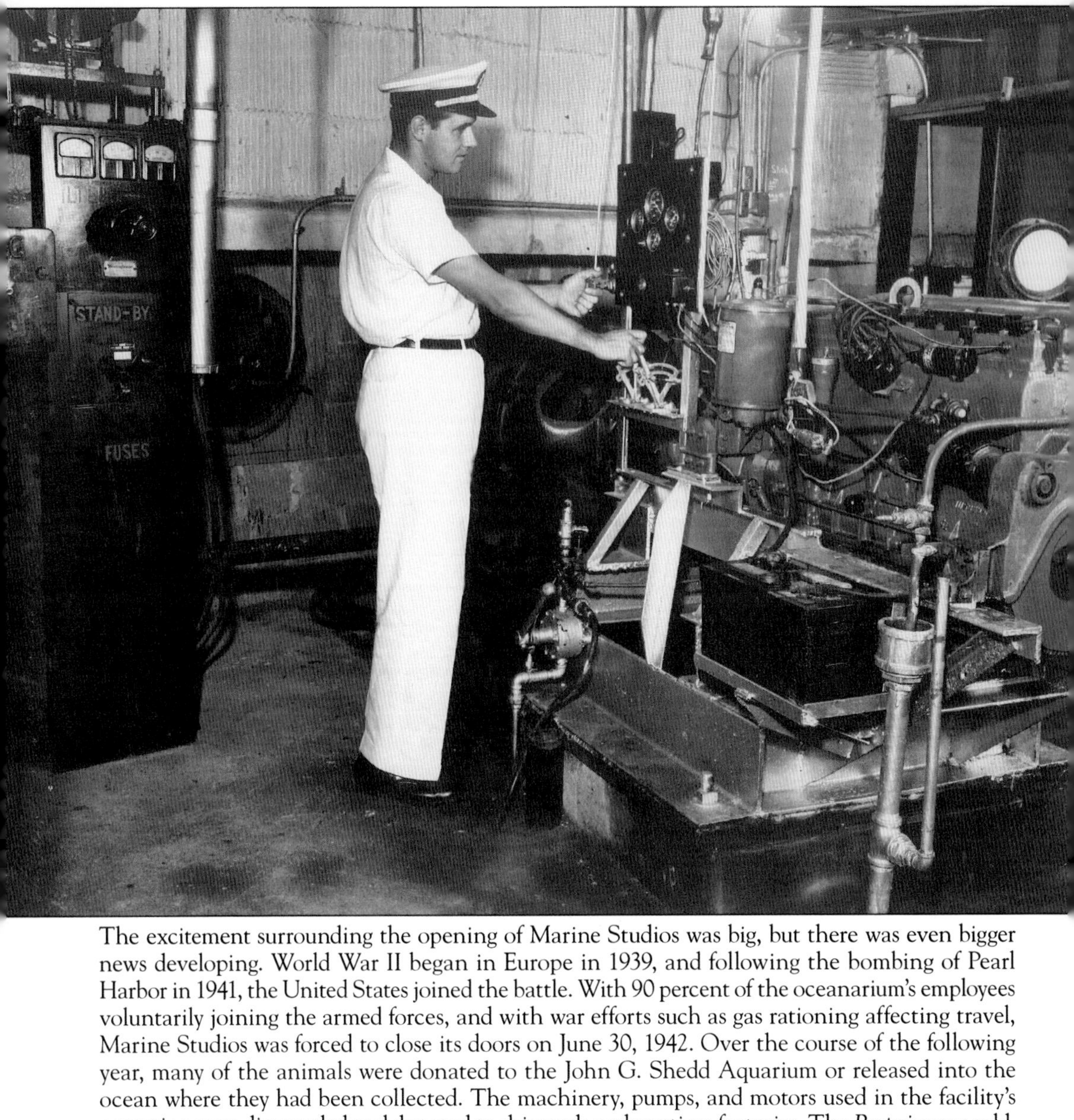

The excitement surrounding the opening of Marine Studios was big, but there was even bigger news developing. World War II began in Europe in 1939, and following the bombing of Pearl Harbor in 1941, the United States joined the battle. With 90 percent of the oceanarium's employees voluntarily joining the armed forces, and with war efforts such as gas rationing affecting travel, Marine Studios was forced to close its doors on June 30, 1942. Over the course of the following year, many of the animals were donated to the John G. Shedd Aquarium or released into the ocean where they had been collected. The machinery, pumps, and motors used in the facility's operation were dismantled and donated to shipyards and wartime factories. The *Porpoise* was sold, and the buildings themselves were turned over to the Coast Guard for national security purposes, with only one employee, Jo VanLandingham, remaining as caretaker of the property. Both Tolstoy and Whitney played vital roles in the American war effort, but when the war ended on VJ Day, August 15, 1945, their hearts and minds were back at Marine Studios preparing for reopening.

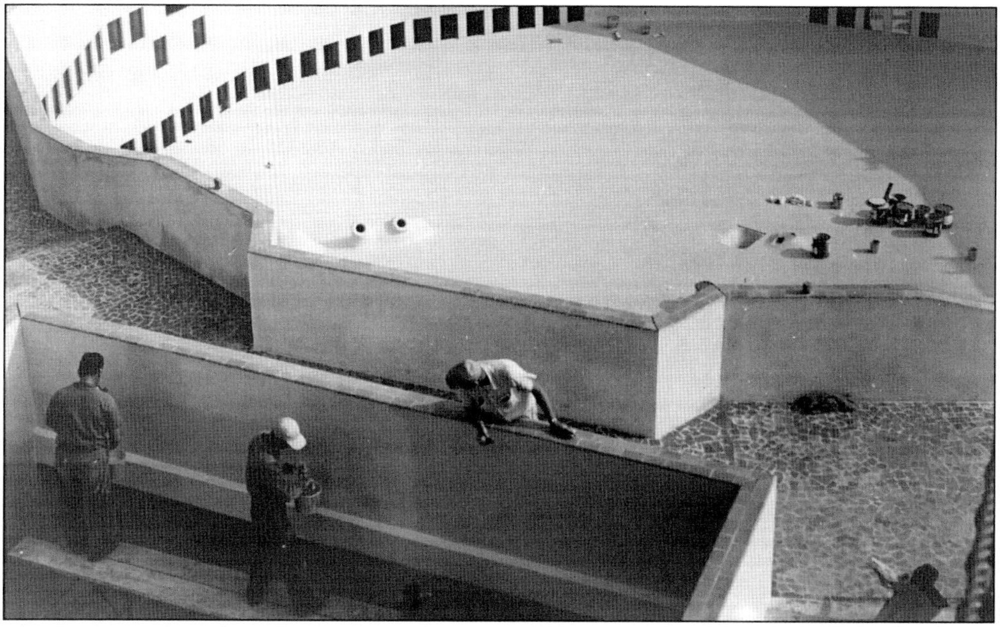

There was much to be done at Marine Studios before the reopening following World War II. Faced with labor and materials shortages, it took about six months to prepare the facility. All of the buildings were renovated and repainted, the public spaces were redecorated, the pumps were replaced, and the gardens were relandscaped. One hundred additional portholes were installed, and each one received new glass fitted with rubber gaskets to prevent leakage. Rock work and shipwrecks were added to the pool floors. The cost was $250,000, about half of the original construction cost. The animal collection had to be restocked, but the collecting boat had been sold. As always, the Marine Studios staff was up for the challenge, and work began immediately.

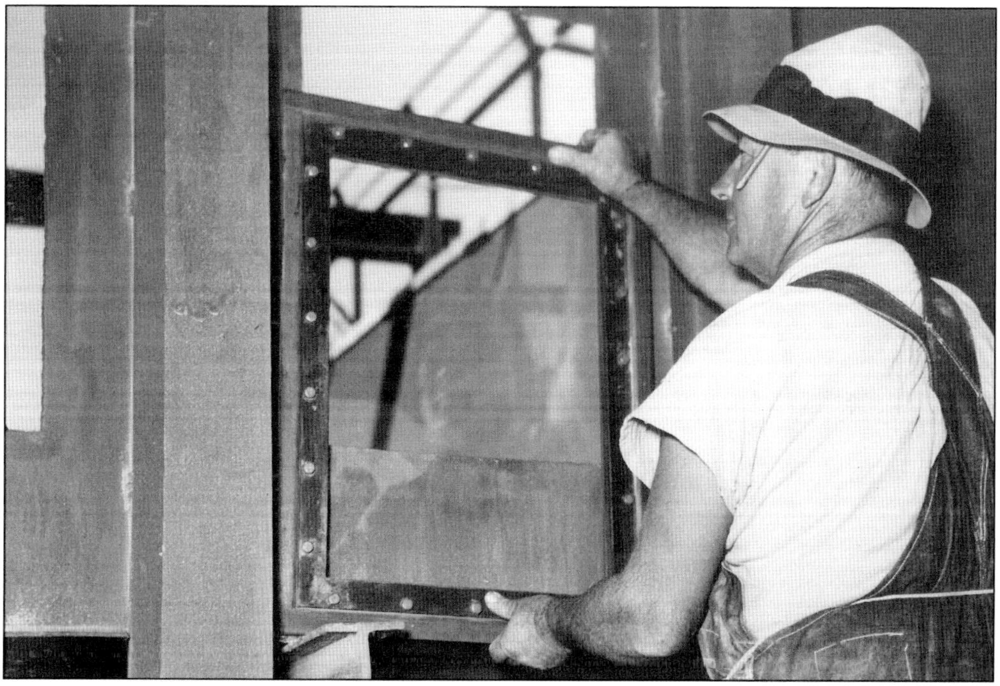

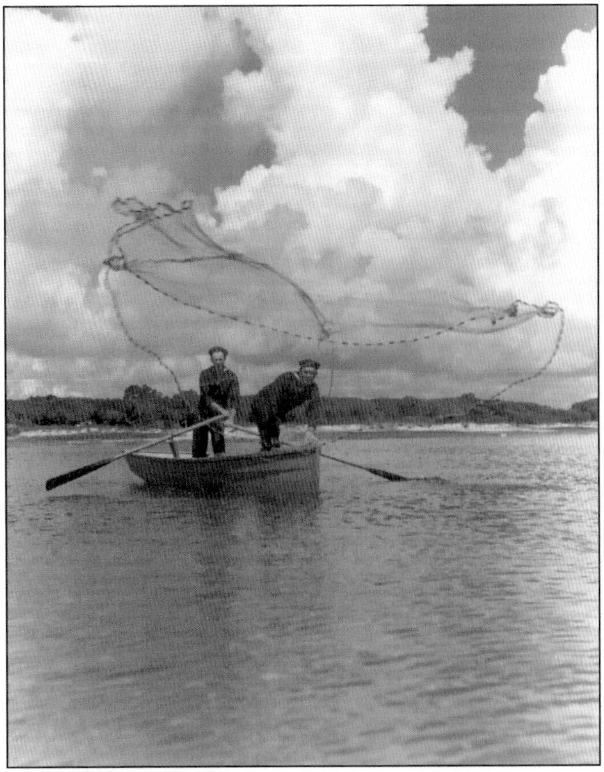

The U.S. Department of the Interior declared that the overall damage to the oceans from the war was so extreme that it might take years for populations of birds, fish, and sea mammals to recover. Destruction of the marine habitat from submarine torpedoes and the death or dispersal of the animals due to bombing, burning oil, and depth changes was estimated to be tremendous. Despite these dire predictions, Marine Studios began successfully collecting again in February 1946. Capt. W.B. Gray was hired, using his 38-foot boat, the *Beau Gregory*, to restock the oceanarium. Gray, a world-class fisherman, discovered more than 100 new species of fish, having more than two dozen of them named after him over his lifetime.

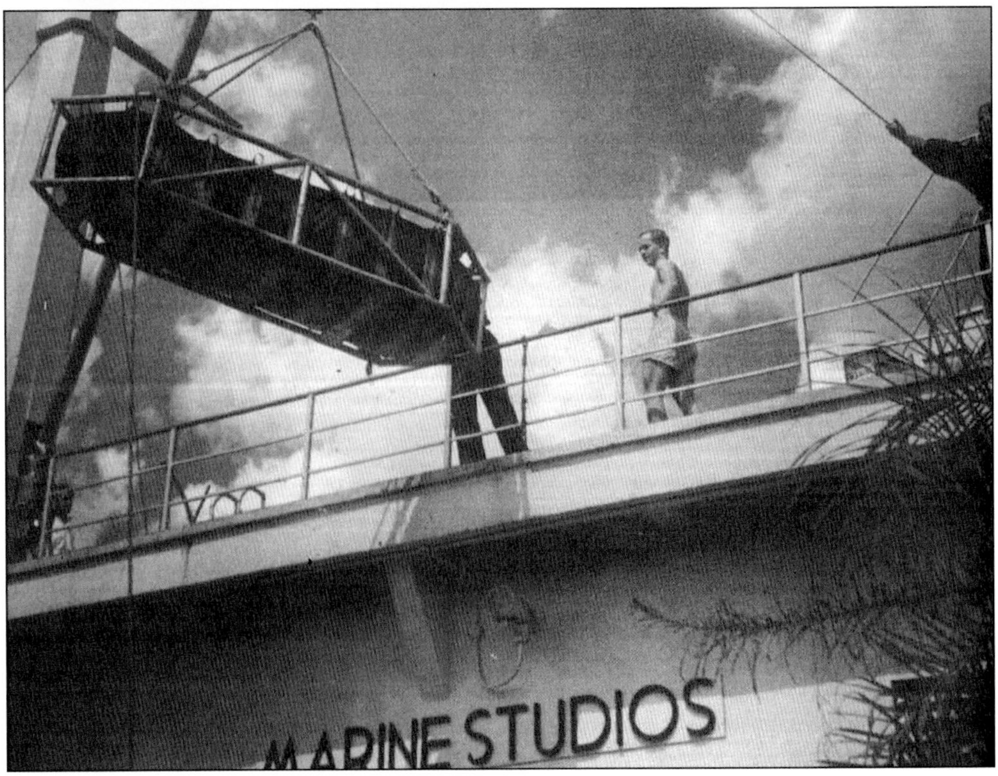

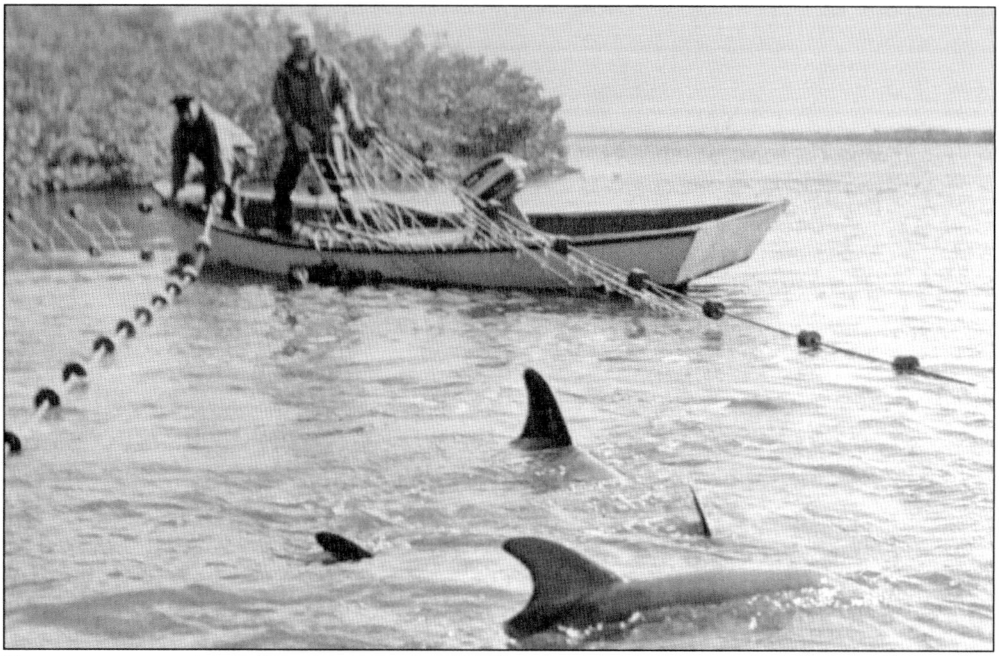

To assist in restocking the oceanariums, the Florida State Board of Conservation waived six regulations to permit Marine Studios to collect fish for exhibition and scientific research, due, in part, to their exemplary history of conservation and educational excellence. Marine Studios collectors were allowed to take speckled trout and mullet during closed seasons, the prohibition on the catch and exhibition of tarpon was lifted, and authorization was given to use crawfish for bait. Additionally, Marine Studios was granted permission to use collection equipment recently forbidden, including purse seine nets, gill nets, and stretched-mesh nets. The state's willingness to contribute to the success of the Marine Studios collection was unprecedented and a true testament to the professionalism of the operation.

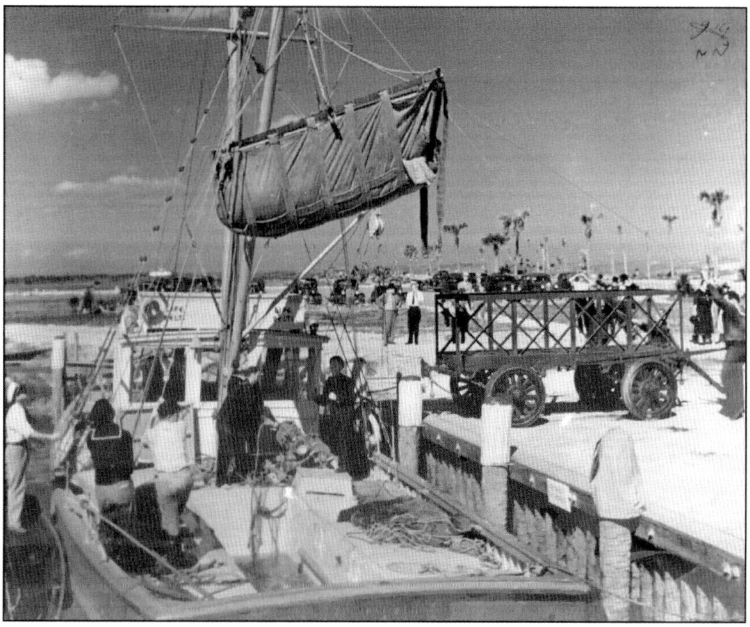

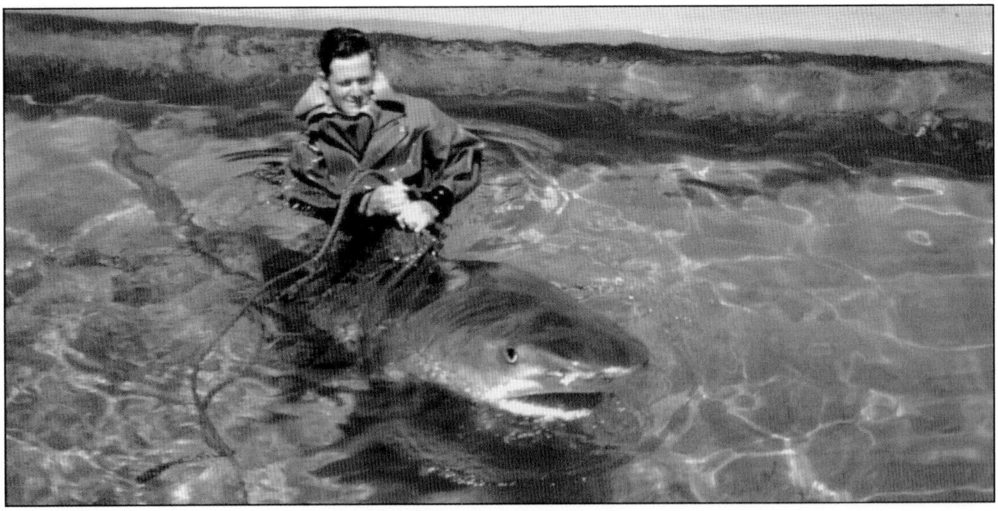

While Capt. W.B. Gray was responsible for restocking the oceanarium after the war, it was curator Arthur McBride's vast understanding of the natural history of marine species and his intelligent observations that were extremely beneficial, leading to many innovative discoveries and improvements. He greatly increased the survival rate of large sharks by developing new ways to circulate seawater in the shipboard holding pools, increasing the amount of oxygen the sharks could absorb through their gills while being transported back to Marine Studios. He found that the use of anesthesia was an unnecessary precaution that prolonged recovery time and quickly discontinued its use. He did, however, preserve the process of shark walking, with brave divers manually increasing water flow and, therefore, oxygen to the gills, until the sharks were ready to swim on their own.

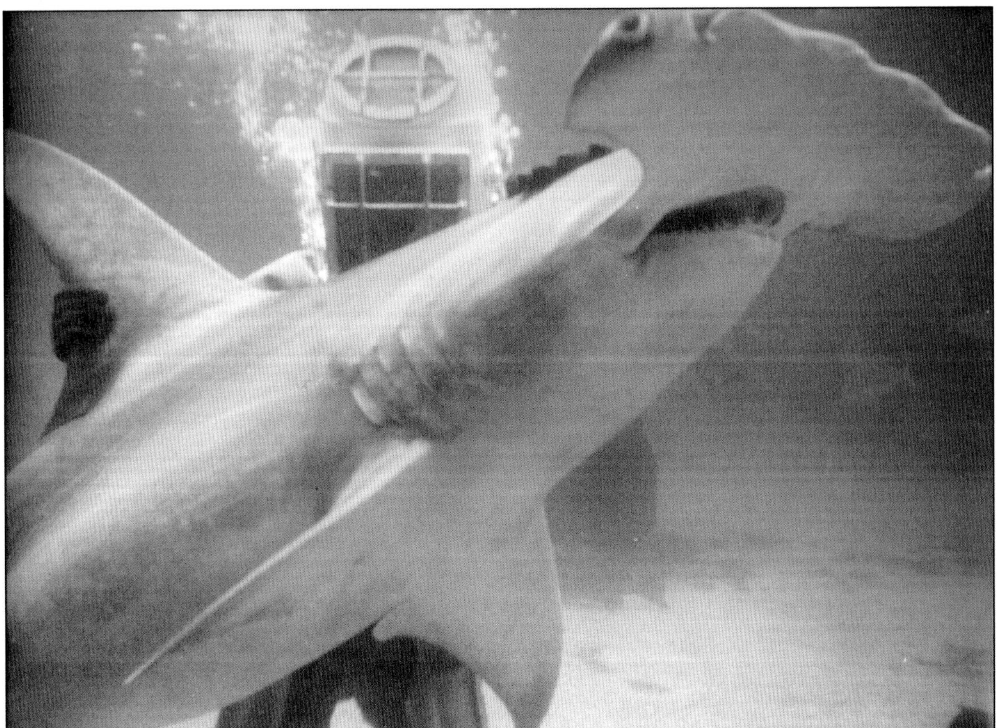

There was little trouble locating the specimens needed to restock the oceanarium. The *Beau Gregory* fished many of the tropical locations surrounding Florida. Assisted by two skiffs, the *Dolphin* and the *Penguin*, the collection team delivered three truckloads of fish per week from shipboard to shore. While both the circular and rectangular oceanariums were being renovated and readied, collected fish, sharks, and dolphins were housed in a natural lagoon directly across the street from the facility. With construction complete, four bottlenose dolphins and a spotted dolphin were moved from the lagoon to the circular pool. Mary Rolleston, wife of general manager Bill Rolleston, painted a No. 1 on the side of the first dolphin, the first specimen, to be placed in the renovated facility. Four newsreel cameramen were present to capture the excitement of Marine Studios' reopening preparations.

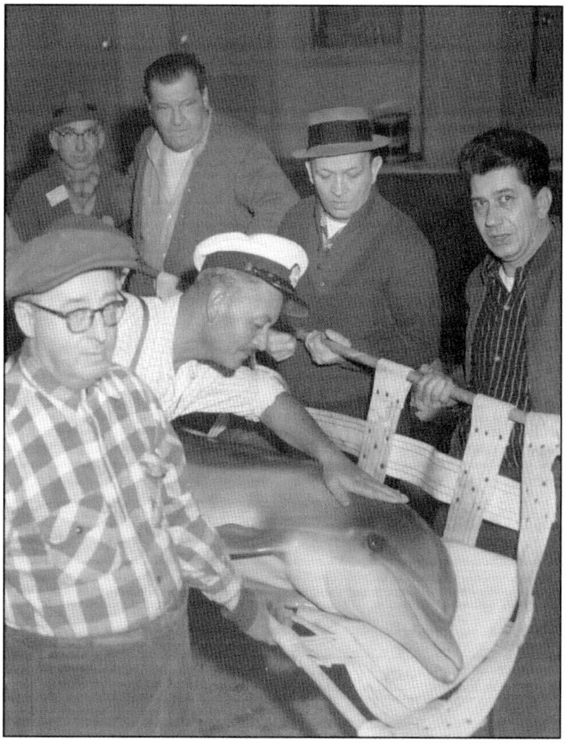

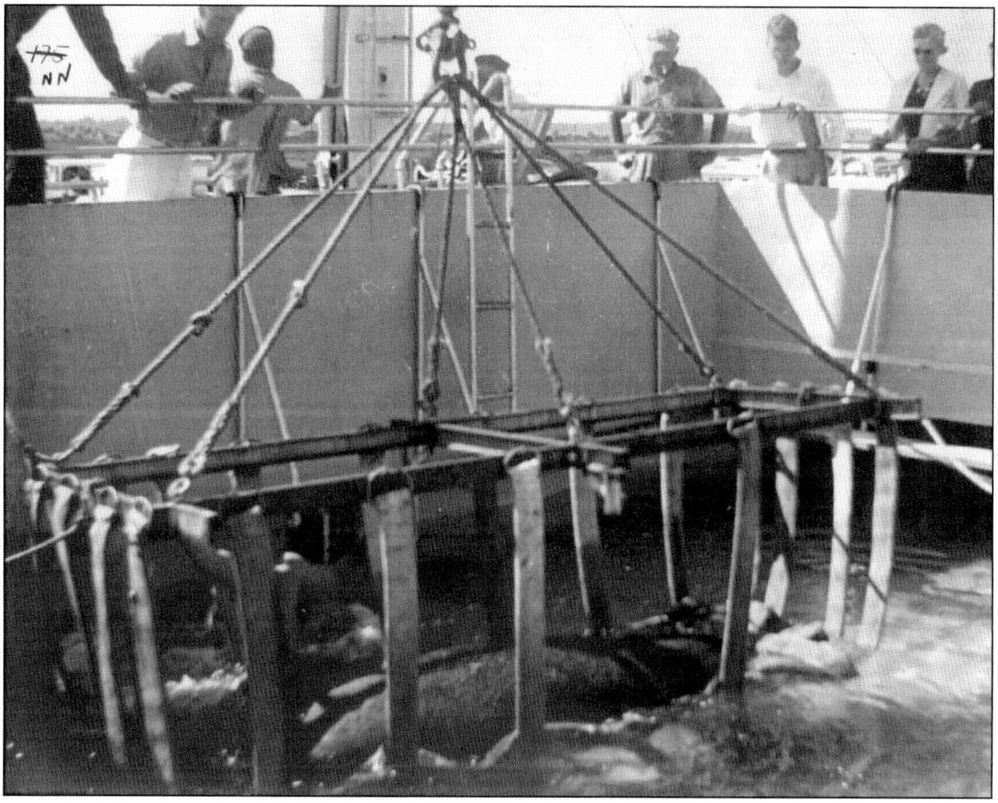

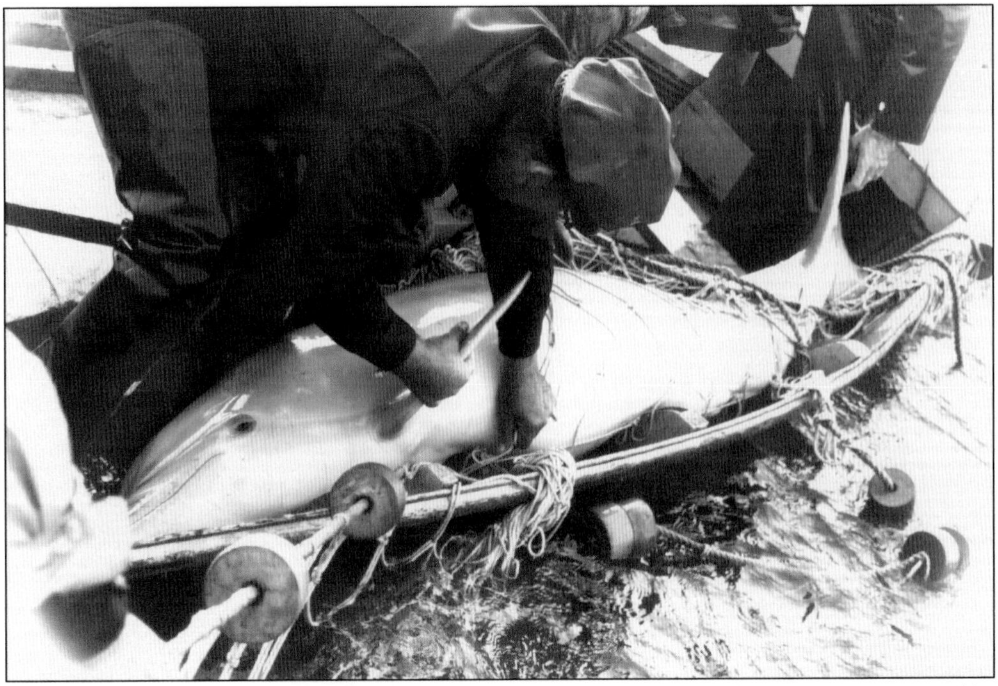

One of the most popular newsreel films in 1946 featured the Marine Studios reopening. Created by Grantland Rice Sportslights, the short film featured innovations in the collection and transport of specimens, including assistance from unusual equipment such as airplanes and amphibious ducks. This story about specimen collection followed an earlier film, *Let 'Em Go Alive*, documenting the specimens' release in 1942, following the close of the facility.

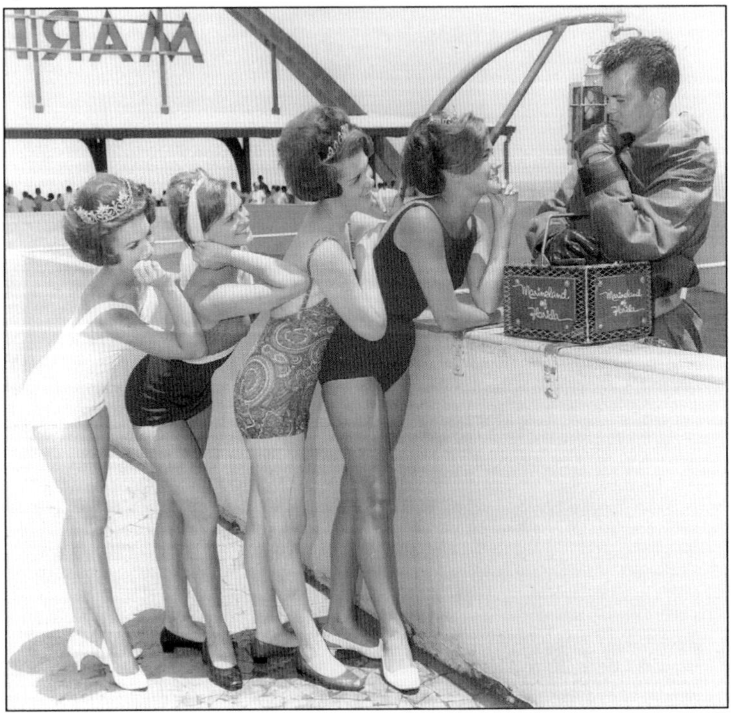

Newsreels were a major source of news, entertainment, and current events. As television increased in popularity in the 1950s, so did interest in Marine Studios. An early local television series titled *Nature Notes* was developed for WESH-TV Daytona in 1957 to highlight the behind-the-scenes work involved in the collection, acclimation, and care of the animals. The staff marveled at the attention, fortunate to have such interesting jobs.

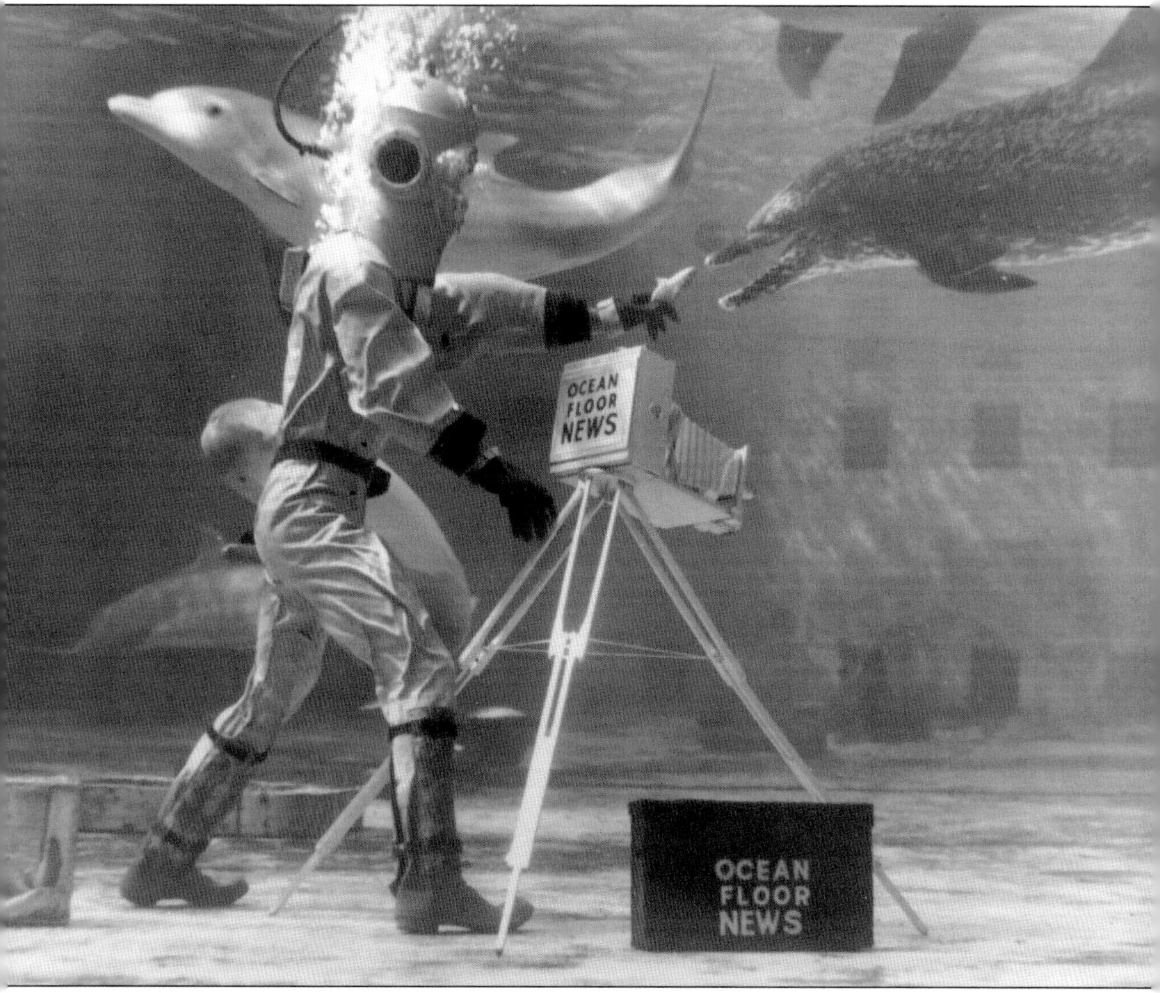

Amazingly, Marine Studios reopened on June 1, 1946, less than one year after World War II ended. Bill Rolleston, the general manager who had overseen the facility's closure, had the honor of orchestrating the reopening. Ceremonies included special presentations by the director of the Fish and Wildlife Service and the secretary of the Florida Press Association. For 30 minutes, a live statewide radio broadcast took place between a diver on the bottom of the circular oceanarium and an announcer on the top deck. The *St. Augustine Record*, which dedicated its entire issue to the reopening, was dropped from circling airplanes to promote the event. Admission was free to everyone, and over a three-day period, 6,000 people visited, increasing traffic on Highway A1A by 110 percent. Suspending operation of the complex during wartime was not the end of Marine Studios. In fact, just as the original opening came on the heels of the Great Depression, the reopening came at the end of the war, just when the country needed it most.

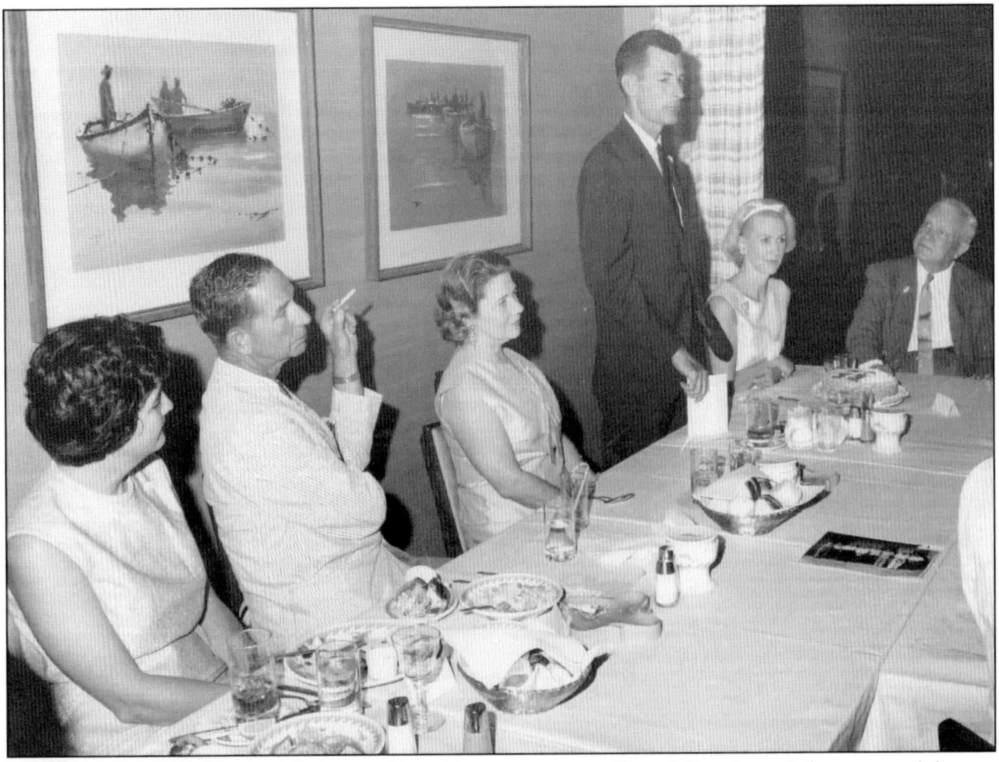

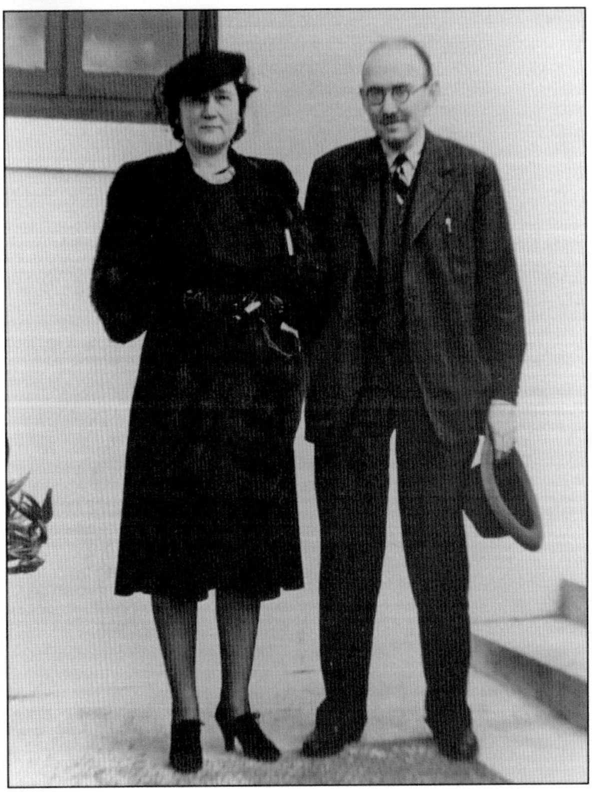

In celebration of the successful grand reopening, a dinner was held at Marine Studios' Dolphin Restaurant, featuring singers Peggy Schaaf and Elizabeth Updike from the Eleanor Searle Whitney School of Music. The glee club of Florida's Normal and Industrial College in St. Augustine also took part in the celebration. About 100 invited guests attended the dinner, which also marked the Dolphin Restaurant's seventh anniversary. Included among the guests was acclaimed author Marjorie Kinnan Rawlings whose husband, Norton Baskin, managed the restaurant as well as the Penguin Bar. Rawlings and Baskin (shown here) were married in 1941, but shared only a short 12-year life together, ending with her death in 1953. Baskin lived to be 95 years old, never remarried, maintained the Rawlings estate, and was generally reported to be the keeper of the flame for Rawlings's literary body of work.

Two

EDUCATORS

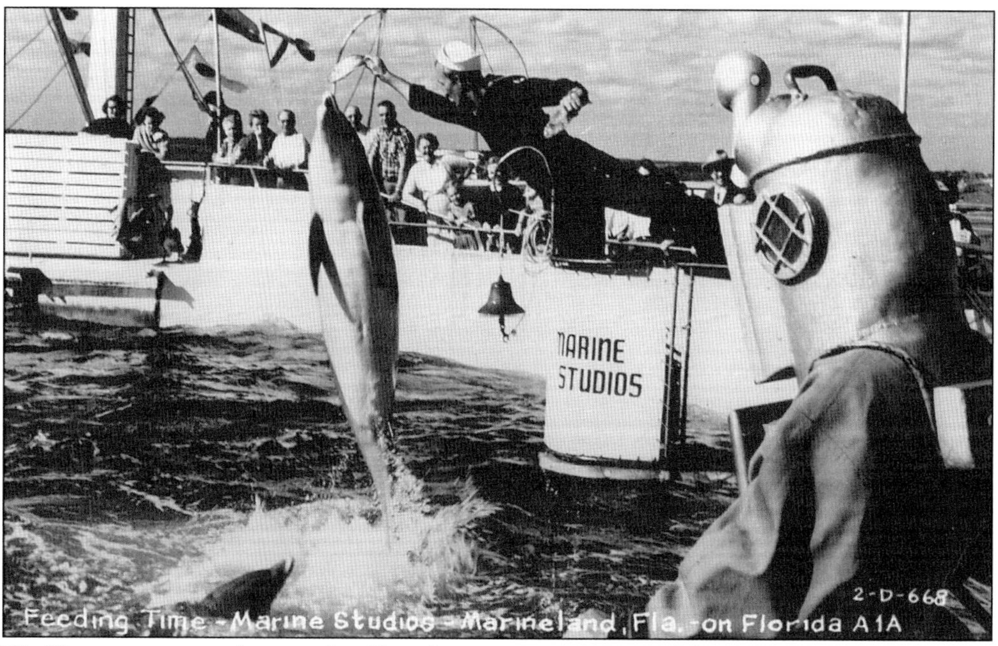

Thrilling interactions between hard-hat divers, jumpmasters, and their amazing marine animals were attractive and entertaining, providing endless opportunities for guests to experience and appreciate ocean life from a valuable firsthand perspective. Whether observing an underwater feeding through porthole windows or cheering as the dolphins soared to new heights, people flocked to Marine Studios for the sheer adventure but left with a better understanding of life beneath the sea.

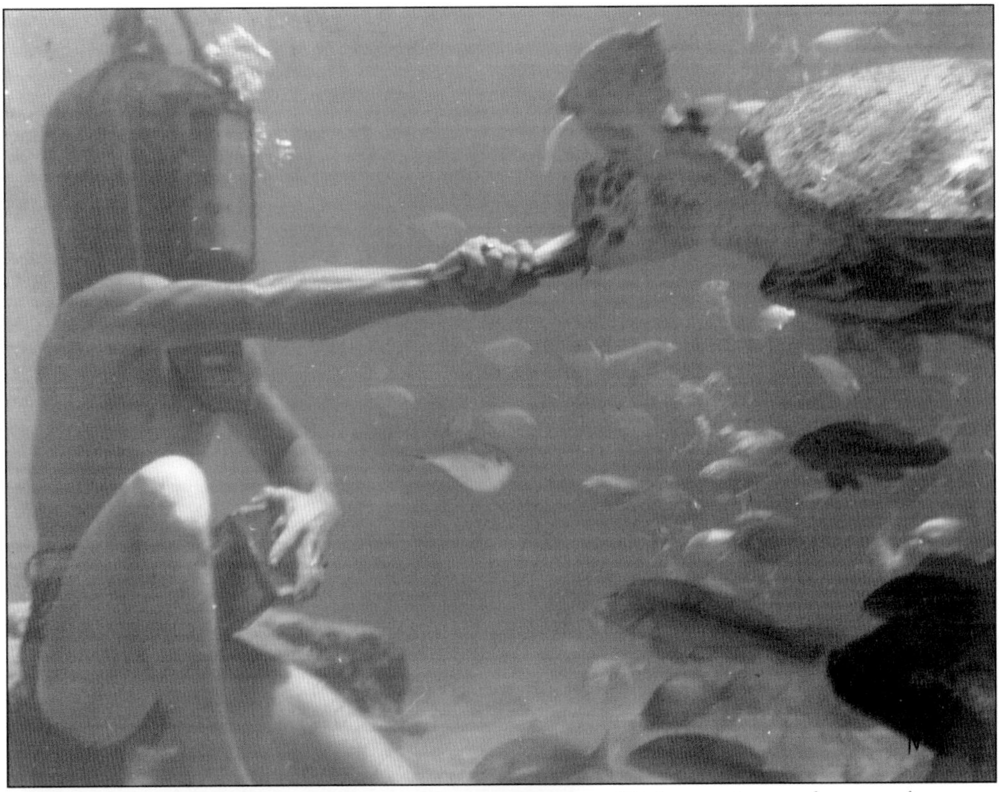

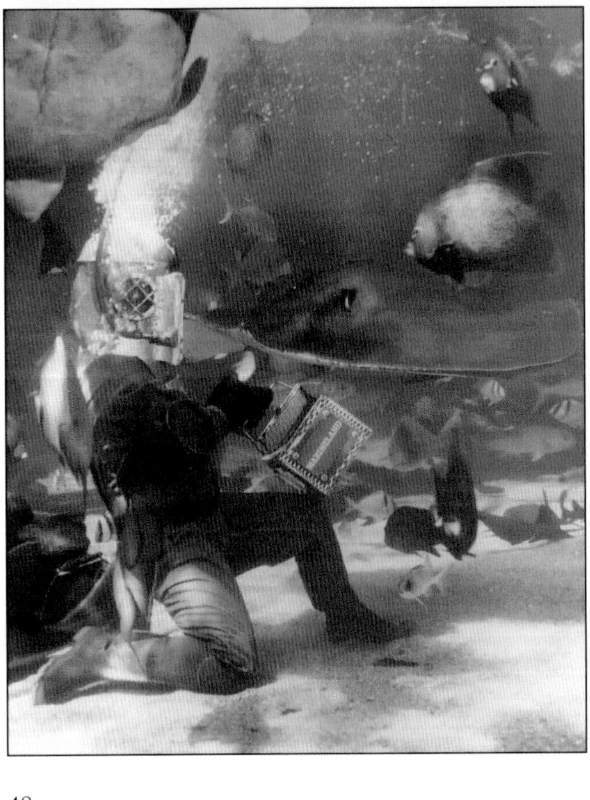

Diving equipment has greatly improved since Marine Studios first opened, but the early hard-hat diving helmets, although cumbersome, were highly functional and served their purpose for the divers, who conducted daily underwater feedings in the oceanarium. The divers carried no air supply. Tethered directly to the surface by a hose, they were fully dependent on their colleagues to provide air from on deck. The copper helmets' considerable weight assisted divers in sinking to the bottom and, depending on the buoyancy needs of the diver, might have also been paired with lead-weighted boots. To surface, the diver could divert breathing air into the dive suit, inflating it and thereby creating the buoyancy needed to exit the pool. The helmets rested directly on the diver's shoulders and could be worn with or without a dive suit.

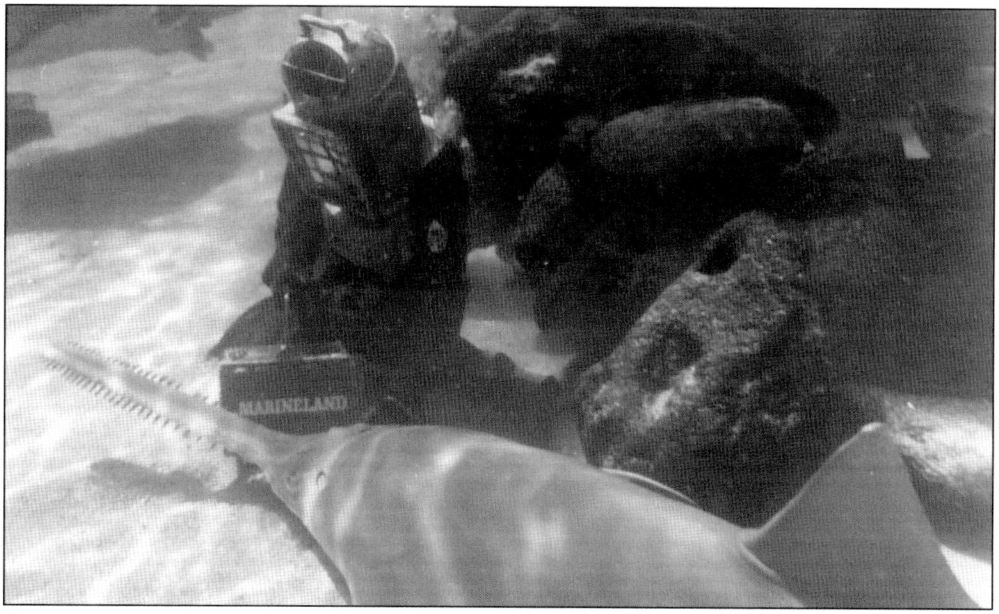

The suits worn by these early divers were manufactured from wool, which has extraordinary insulating capabilities in air, but was found to have poor insulating qualities when used underwater. In 1940, after the coldest winter in Florida's recorded history, Southern Air Conditioning Corporation of Jacksonville installed a General Electric automatic air-conditioning and heating system to the 450,000-gallon rectangular oceanarium, now referred to as the tropical tank. The system, utilizing five large oil furnaces, allowed the pool to be held at 70 degrees Fahrenheit at all times. The heat exchange took place in the filter banks, and thermostats in the pipeline regulated the temperature. It was the largest heated body of water in the world and was greatly appreciated by the fish and divers alike.

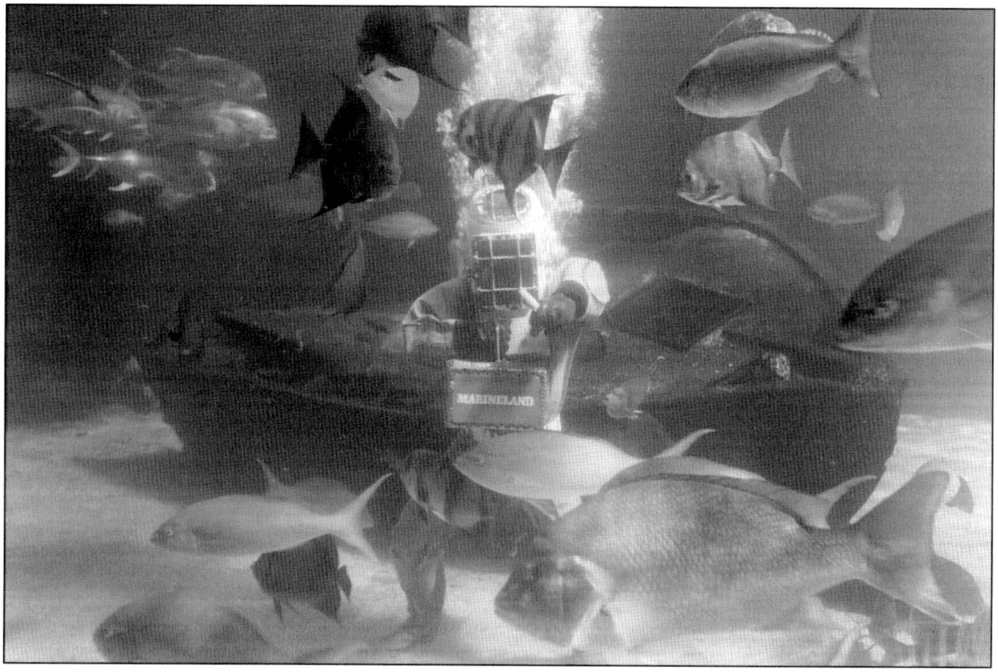

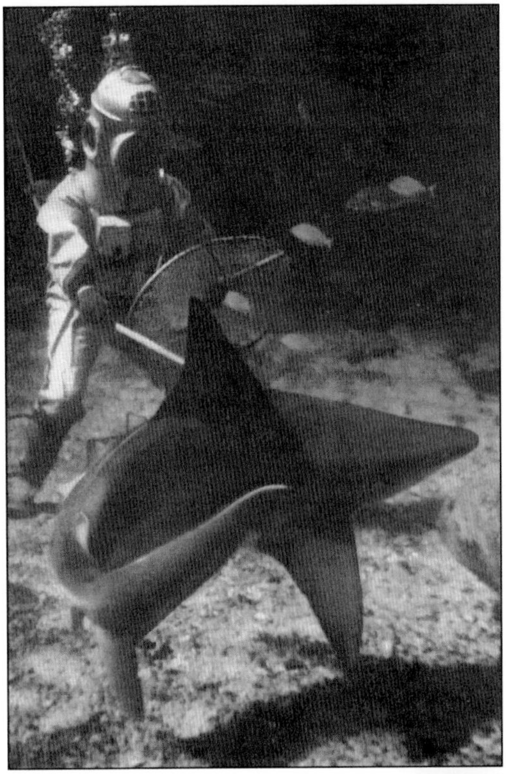

The hard-hat divers were the earliest featured attraction at Marine Studios, descending 15 feet below the surface of the water to directly interact with the seemingly deadly specimens living there. A red flag flown on the topside mast of the oceanarium signified "diver down," causing excited crowds to rush to the portholes, the perfect vantage point from which to watch a diver fend off a man-eating shark or a menacing moray eel. The red flag, known as the Navy bravo flag, is universally flown on ships to signify danger. Incorporating this red warning color, the dive flag as it is known today was invented 15 years later in 1953.

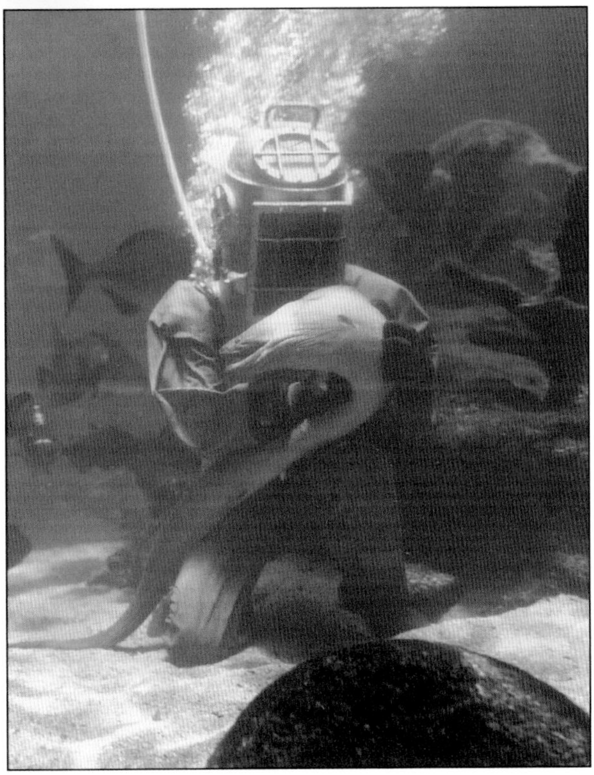

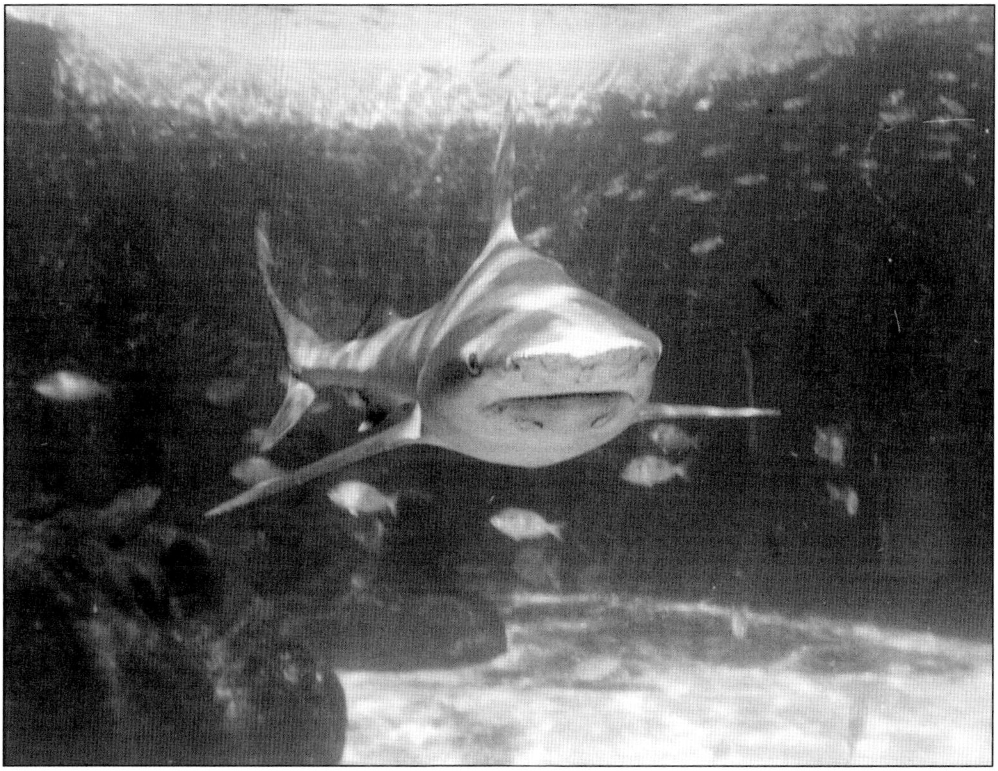

Preconceived notions about the dangerous and calculating nature of the large predatory sharks were certainly a customer draw, but also led everyone to believe the Marine Studios divers needed full protection, often in the form of weapons to fend off any possibility of attack. When curious sharks began following divers around the bottom of the oceanarium or began biting at hoses, they were immediately returned to the ocean. When the collecting team brought in fearsome tiger sharks in 1940, the divers stopped underwater feedings for the first time until a protective cage could be built. Feedings then continued from the submerged dive cage. But with time came understanding, and open feeding resumed once again when divers realized the tiger sharks were no more aggressive than the other species, regardless of their imposing appearance.

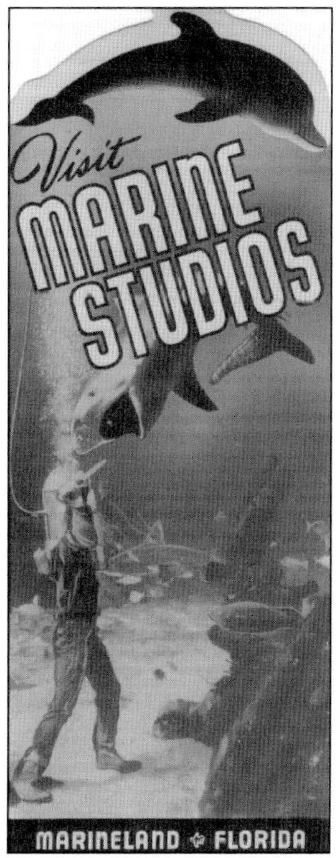

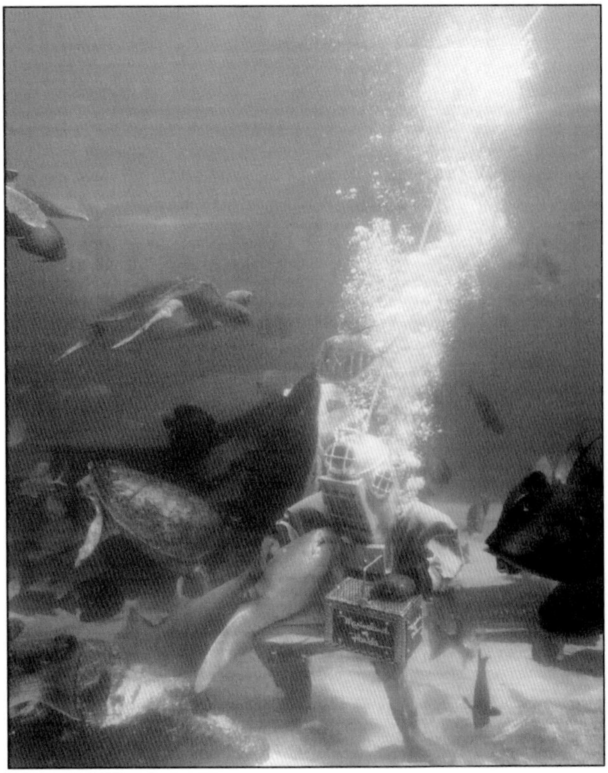

The Marine Studios divers had the unique opportunity and perspective to make important discoveries about sharks, which continue to be one of the most misunderstood animals today. Back in the 1940s, sharks were even more of an anomaly. The divers quickly realized that not all sharks were to be feared. To the contrary, they discovered that nurse sharks were some of the most docile and gentle fish they cared for. Thriving in the oceanarium, nurse sharks were the first to take fish directly from a diver's hand. The divers' daily interactions with sharks sometimes helped to disprove myths, including the long-held myth that sharks must turn upside down to feed. They also made numerous new discoveries, including the realization that sharks have a never-ending supply of teeth that move forward to replace lost ones.

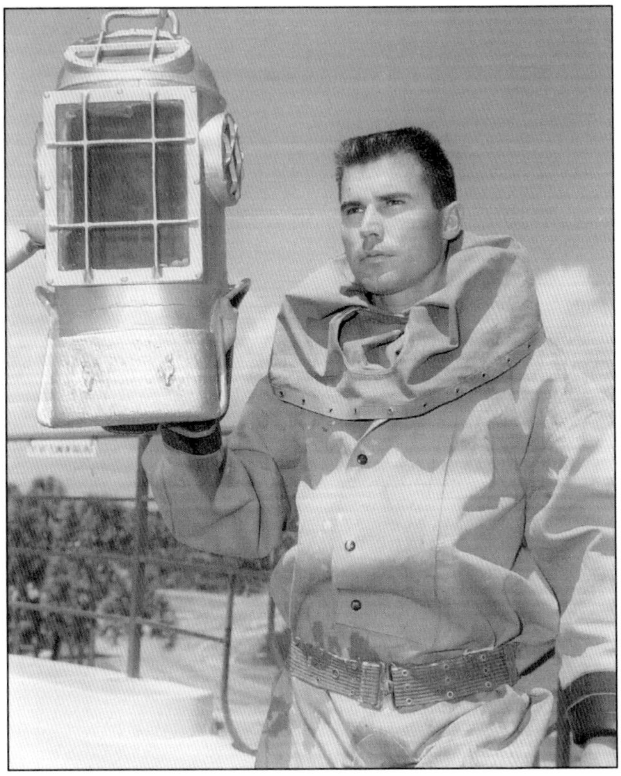

At first, Marine Studios attempted to attract the attention of professional deep-sea divers to take on the role of oceanarium divers. The founders were dumbfounded when each offer of employment was flatly refused; the job appeared much too dangerous, even for professionals. Believing in their vision, it was Marine Studio's general manager Ilia Tolstoy and curator Arthur McBride who became the first divers to enter the oceanarium. Their repeated bravado and diving demonstrations summoned courage while building understanding and acceptance. Soon a long line of local Florida boys were interested in the job, eager for a chance at adventure. These first divers inspired a whole generation of young men who watched from the portholes, transforming the diver position from the most feared to the most sought after job at Marine Studios.

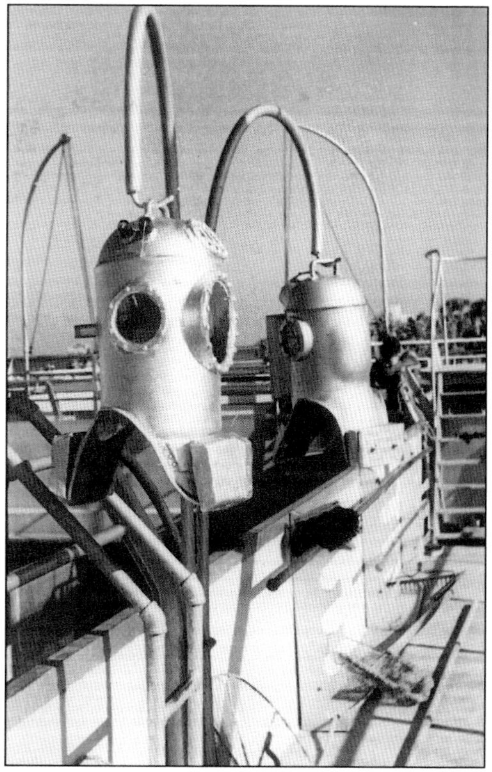

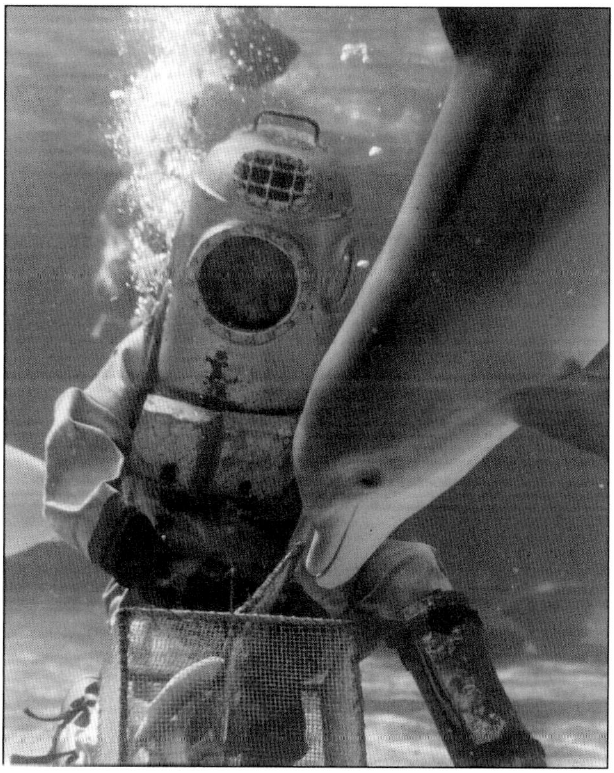

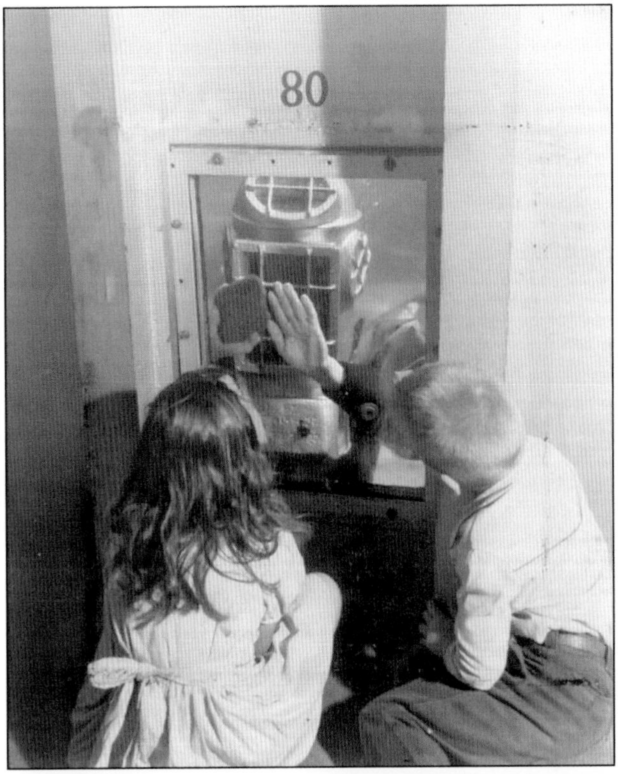

Despite the state-of-the-art filtration system and the copper-treated water, the endless Florida sunshine made controlling algae growth in the pools an unexpected and seemingly endless job for the divers. It included the daily cleaning of hundreds of underwater windows, much to the chagrin of the divers and the delight of the visitors. Vacuuming the sand-and-coquina-rock floors and even scrubbing the algae from the shells of slow-moving sea turtles fell into the diver's job description. This laborious work was necessary to maintain the overall health and well being of the animal collection. Even something as mundane as cleaning became exciting at Marine Studios, when Fox, Pathe, and Hearst Metrotone newsreels featured the activities in a 1940 short film titled *Spring Cleaning.*

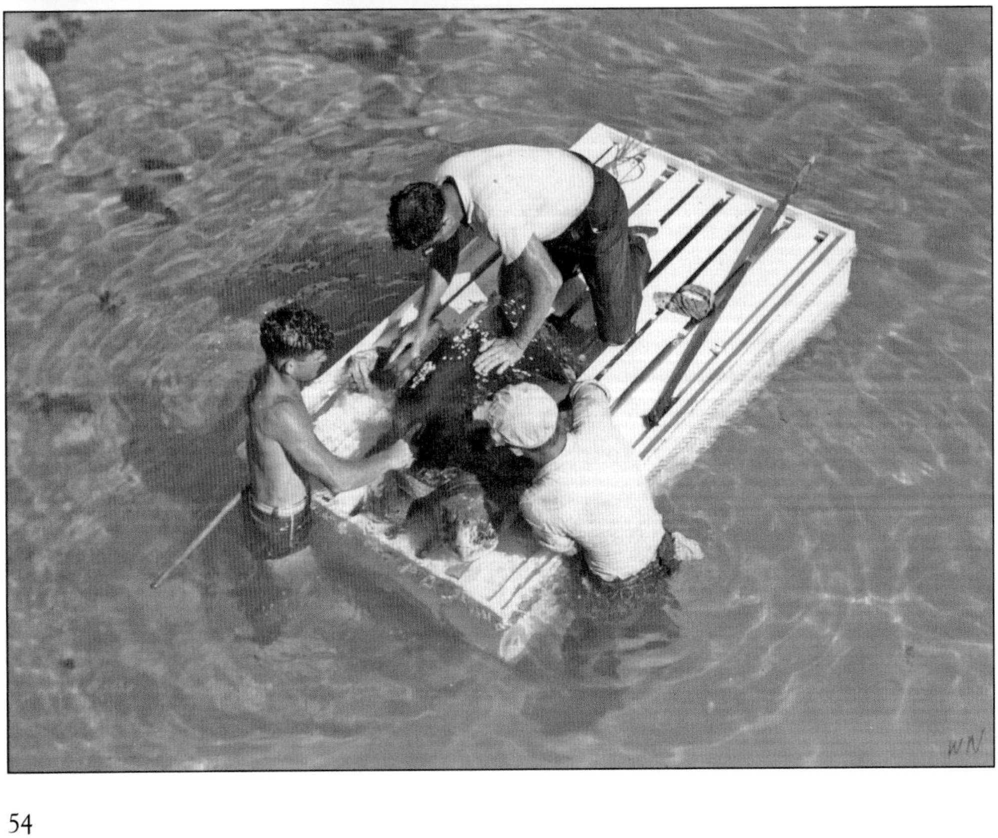

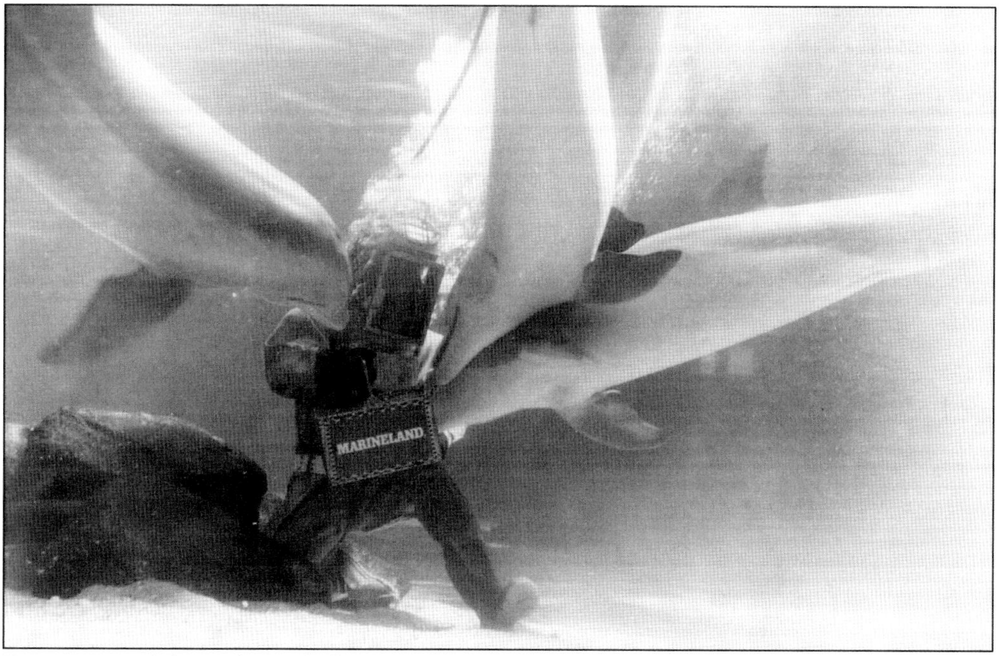

Diver responsibilities also included underwater feeding of the dolphins. Agile, fast, and curious, the dolphins were quick to surround any diver carrying a wire basket filled with fish treats. Although bits of food were offered throughout the day, the animals were fed their main meals outside of public demonstrations such as these. Marine Studios' first dolphins ate approximately 20 pounds of restaurant-quality fish each day. A bell sounding underwater just prior to feedings was initially used to alert the dolphins that mealtime was about to begin. Although the dolphins quickly learned that the gathering crowds signaled feeding time, the use of the bell was continued for its nautical effect and its efficient way of attracting the public. The picture below suggests that the sea turtles learned the meaning of the underwater bell as well.

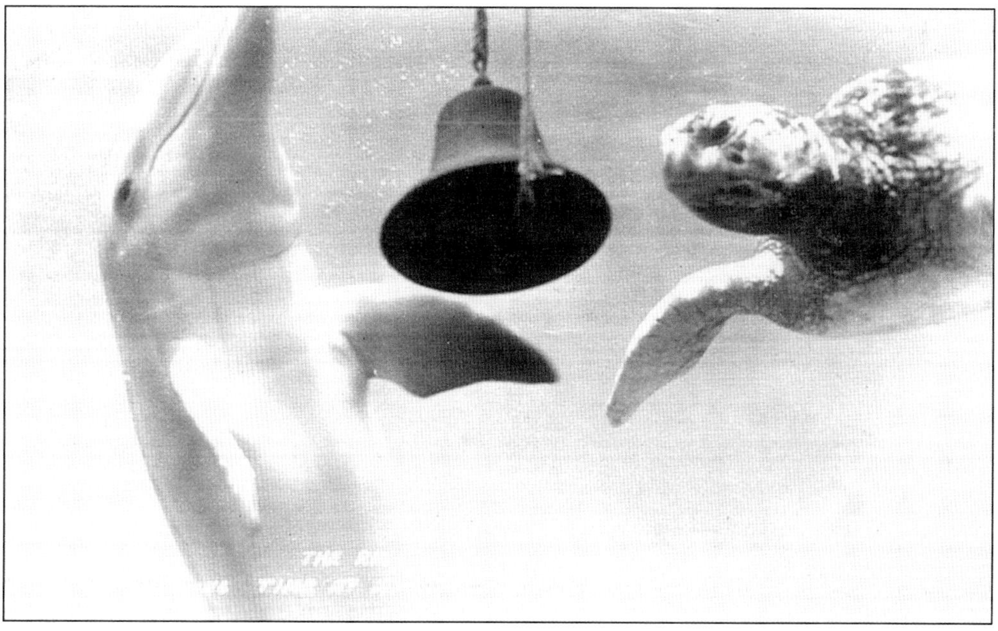

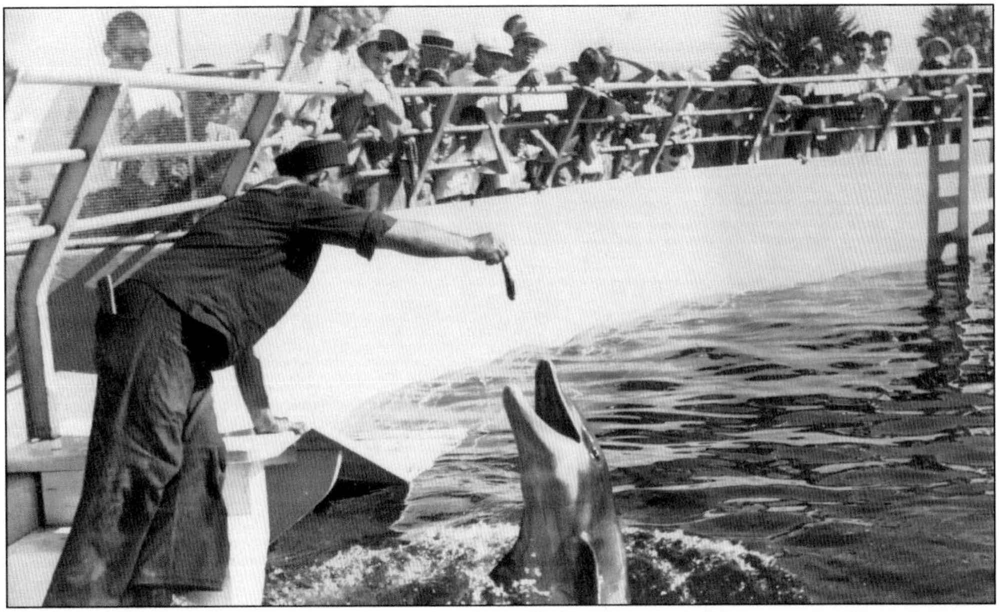

Marine Studios staff was proudly caring for the only bottlenose dolphins in the United States. Tossing fish to them as a means of supplying food progressed to hand feedings at the surface. It was soon evident that by holding a fish slightly above the water, the dolphins would rise up to grasp it. This was the humble beginning of what was to become a signature attraction, the famous jumping dolphins.

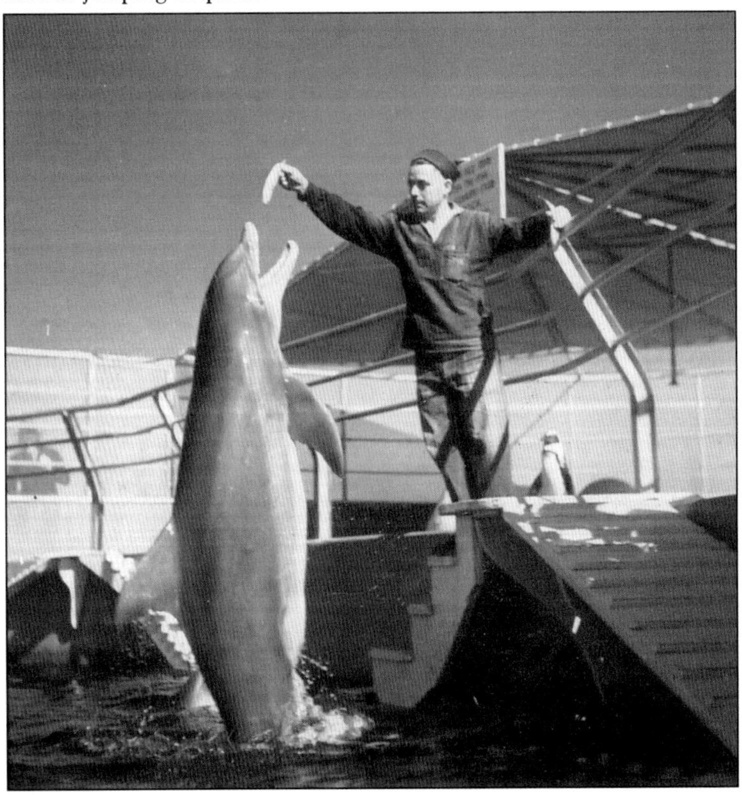

Both competition and mimicry served to produce an entire pool of jumping dolphins, creating the very first dolphin presentation, which became so popular it was performed up to six times each day. The naturally creative and innovative behavior produced by Marine Studios' jumping dolphins was legendary and led to a new staff position: jumpmaster.

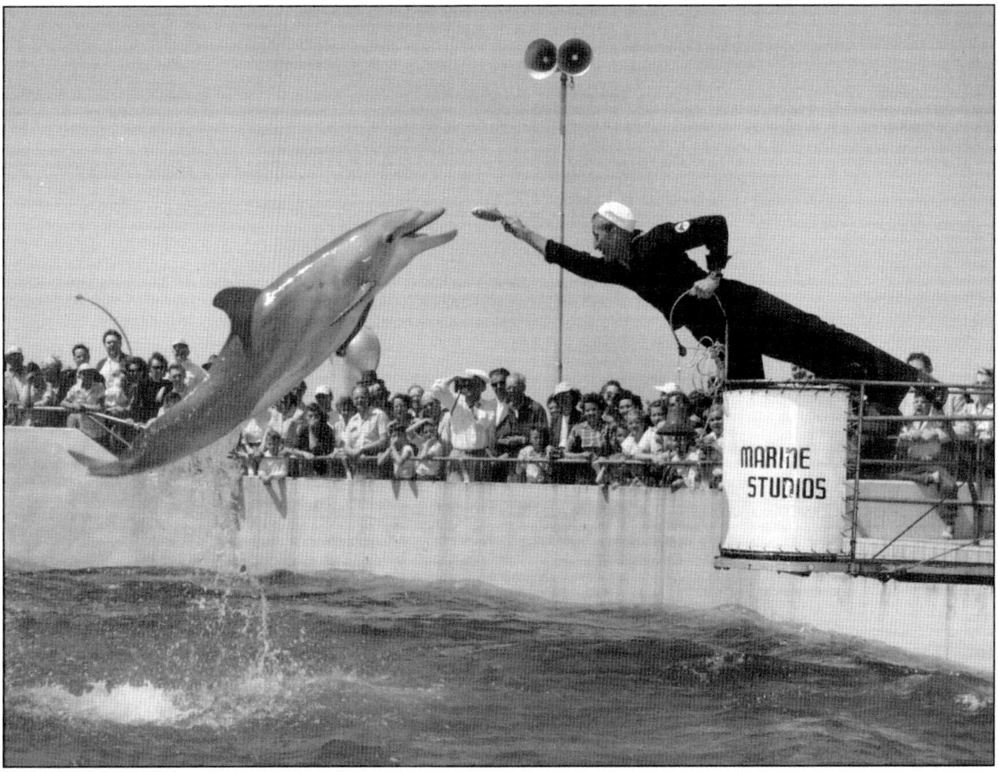

The most famous jumpmaster at Marine Studios was Mitch Lightsey, who was born in Claxton, Georgia, in 1912. He was one of eight children. Initially hired as a diver for the oceanarium in 1941, he moved into the jumpmaster position following the reopening of Marine Studios, where he remained for a lifetime. On March 7, 1948, the *Florida Times Union* declared Lightsey "the most photographed person in America," thanks to the dolphins.

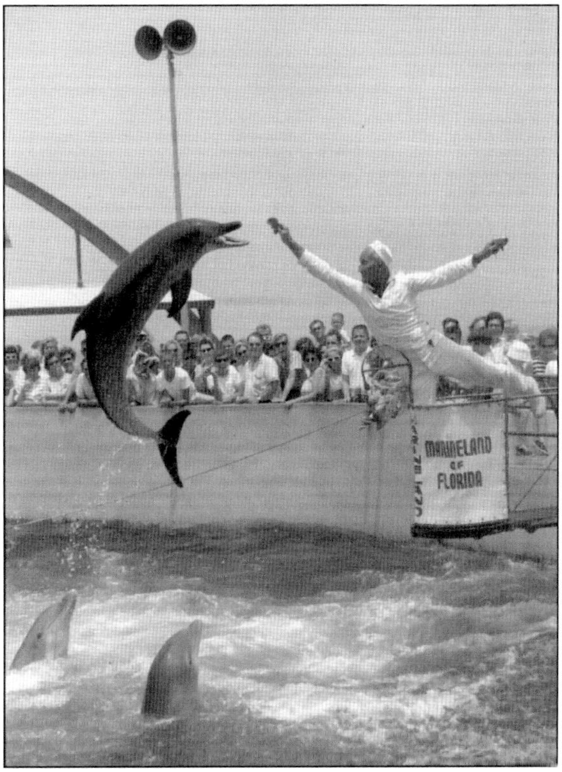

Lightsey had no special training to prepare him for this high-profile job. Before moving to Florida, Mitch owned a bowling alley and poolroom and worked for the Georgia Road Department. His vast skills were learned on the job through hands-on experience. Although he and his wife, Eloise Lee, had no children of their own, Lightsey raised, nurtured, and trained multiple generations of dolphin calves in the circular oceanarium.

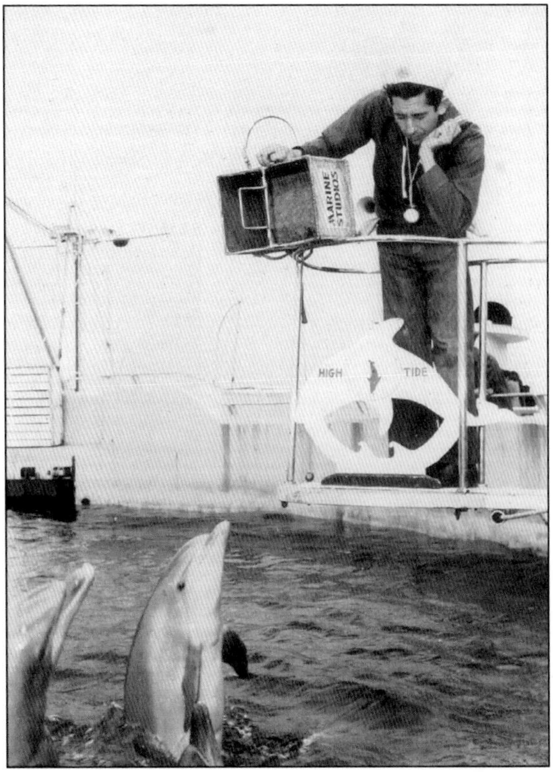

Guests were constantly amazed at the willingness and interest of the dolphins to interact with them. It was frequently observed that the dolphins initiated play with people as much as people did with them, even when there was no food involved. Camera lenses, hats, and glasses that were inadvertently dropped into the pools by careless guests were quickly returned by the dolphins, much to the guests' delight.

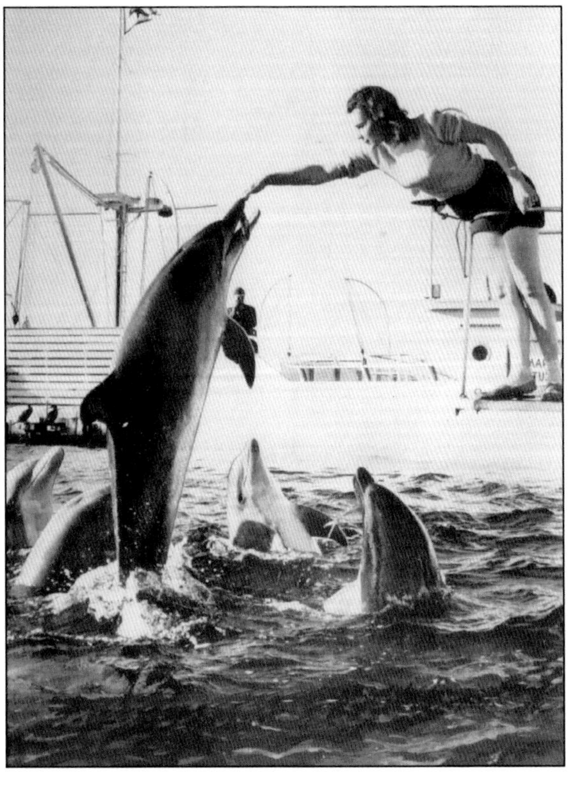

Expanding upon the public's enjoyment of playing with the dolphins, Marine Studios added audience participation to its jumping-dolphin show in 1946. A visitor was chosen to occupy the jumpmaster's feeding platform, encouraging the dolphins to jump up and take a fish from his or her hand. The participating guest received a personalized, signed, and dated jumpmaster certificate as a souvenir of his or her accomplishment.

By 1940, the circular pool exhibited a large population of playful bottlenose dolphins. Although man-made toys were provided, the dolphins would often create games of their own. Bird feathers turned out to be a favorite toy. The dolphins would take the feathers in their mouth to the bottom of the oceanarium and release them in the current of the water jets, where they would soar off and could be chased. The dolphins would also solicit body rubs from the staff, which they appeared to enjoy very much, whether from the jumpmasters at the surface or the divers underwater. As added enrichment, the divers attached brushes to the bottom of the pool, allowing the animals to receive rubs even when the staff was not present.

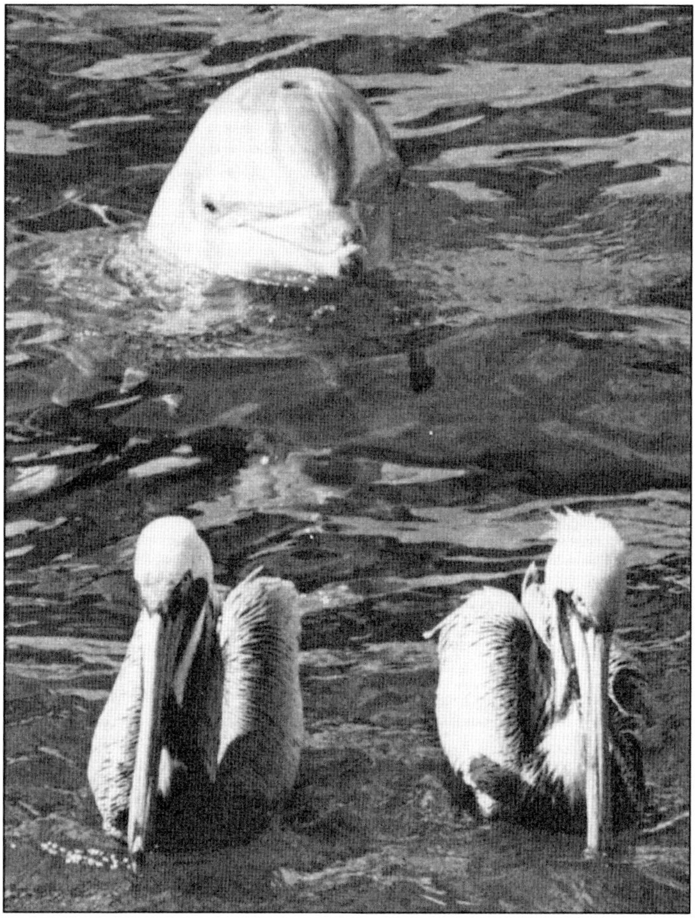

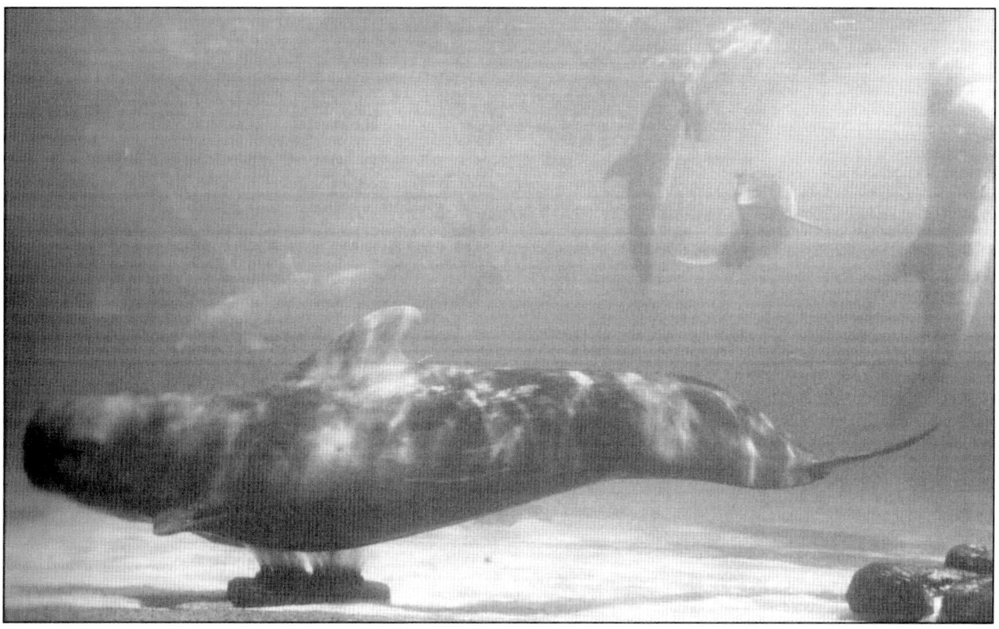

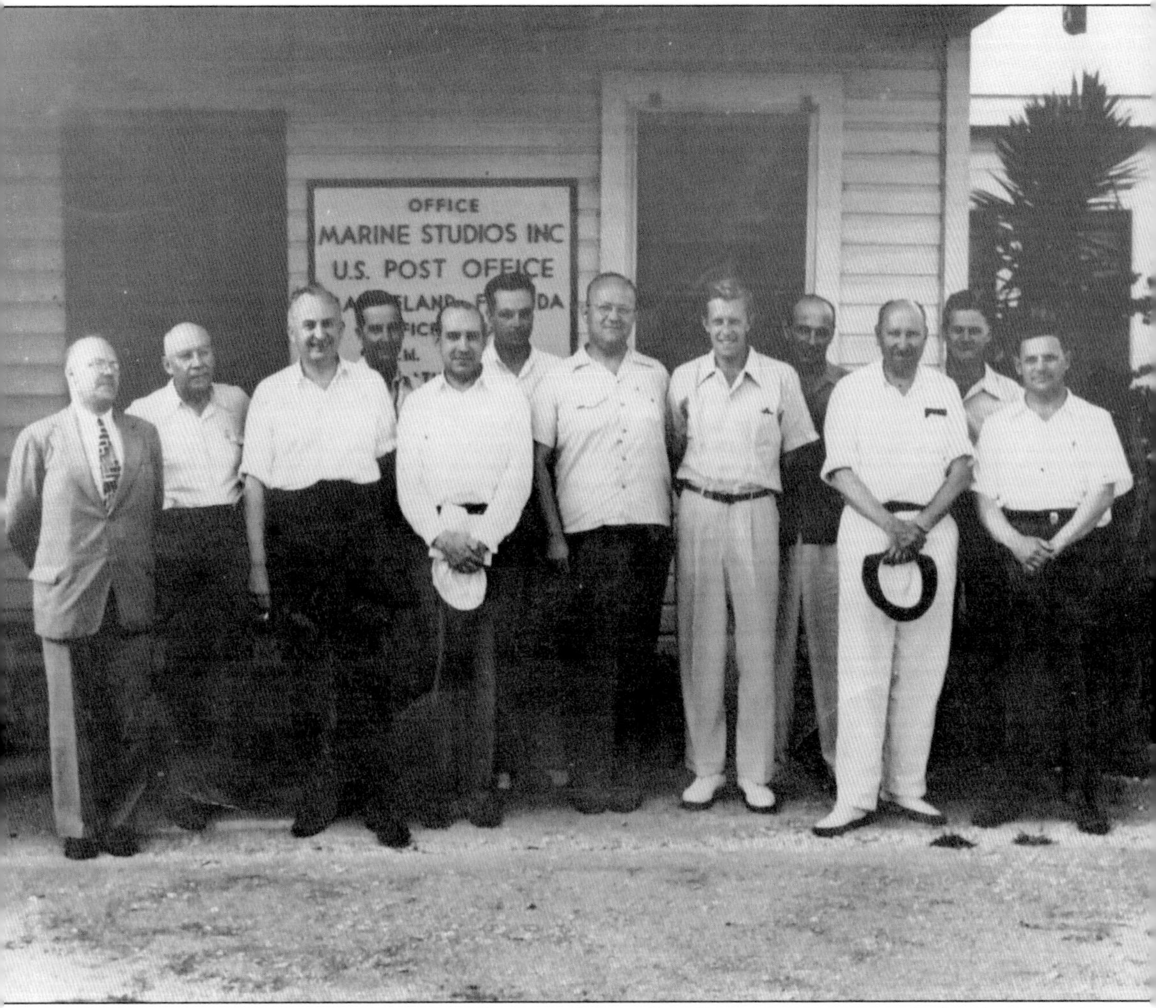

On November 1, 1946, the town of Marineland, Florida, was incorporated. As the newest municipality in Florida, the town's name originated following the construction of Marine Studios. Known to be one of the smallest towns in the state, Marineland boasted only 26 registered voters. In 1950, Marine Studios performed its own comical rendition of a government census in the circular dolphin pool. The point made in the Marine Studios' spoof was clear—there were actually more marine inhabitants of the town than there were people. In fact, the fish outnumbered the humans in the town of Marineland by 2,000 to 1. The town's first mayor was Milton E. Bacon Jr., Marine Studios' director of public relations.

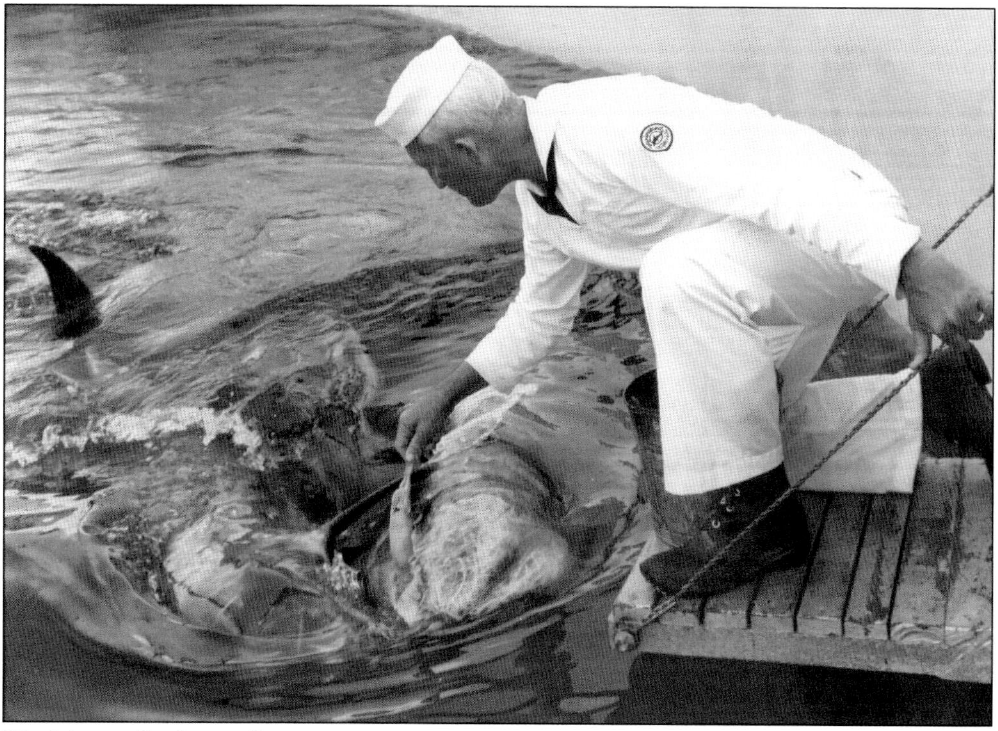

The Marine Studios staff were pioneers and experts in the care and nurturing of ocean animals. When two live Risso's dolphins were stranded on Crescent Beach in the late 1960s, it was the Marine Studios staff who came to their rescue. The two males, sick and injured, were brought directly to the oceanarium for rehabilitation. On a steady diet of about 60 pounds of squid per day, they were nursed back to health, becoming the only Risso's dolphins to be displayed in the Western Hemisphere. This unusual pair was named Elsworth and Dennis, after the men who first reported the stranding event that led to their rescue. At 11 feet long and 650 pounds each, they became a beloved addition to the jump show in the circular pool, performing together in perfect unison.

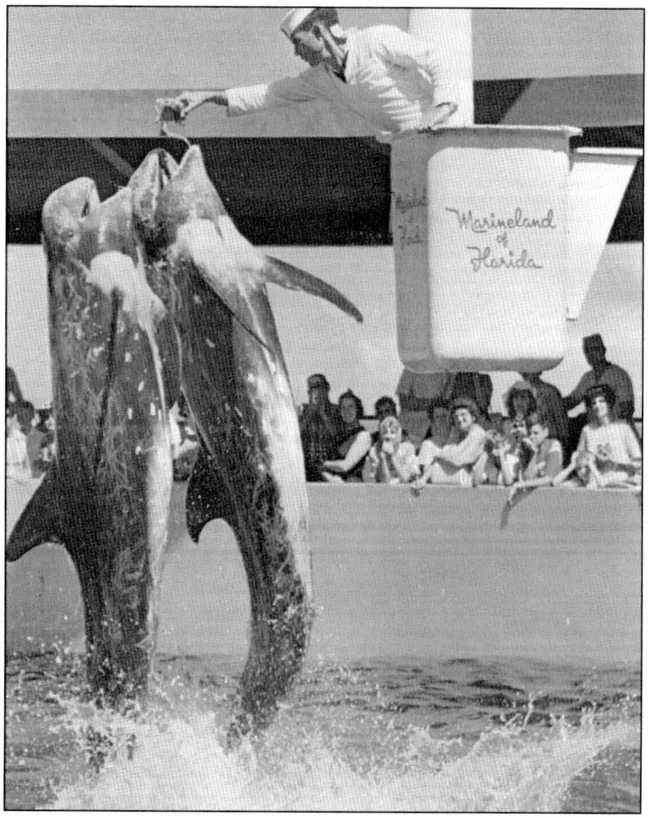

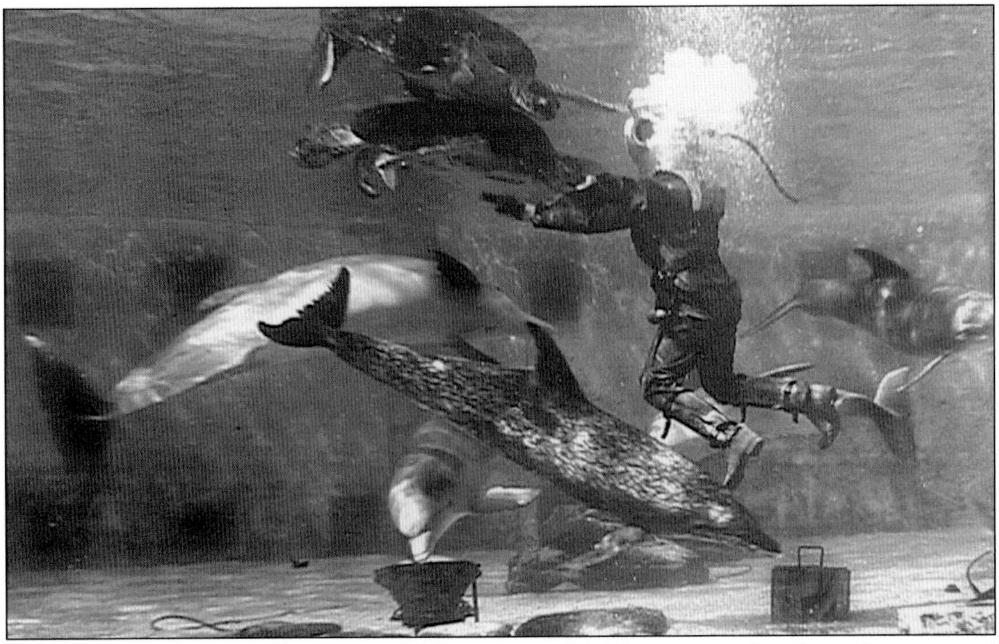

Along with the Risso's and bottlenose dolphins, spotted dolphins, also referred to as long-snouted dolphins, found a home at Marine Studios. Typically found in large numbers in deeper oceanic waters, they had, to date, been unable to thrive in any oceanarium. Under the skilled hands of the Marine Studios staff, spotted dolphins were not only displayed by 1940, but their beauty and grace was presented in daily exhibitions as early as 1951. The staff soon discovered that the spotted dolphins had the ability to jump much higher than their bottlenose cousins. The most famous of the spotted dolphins was Dottie, whose aerial feats thrilled crowds for 10 years, serving as an incredible testament to the expertise and skill of the Marine Studios team.

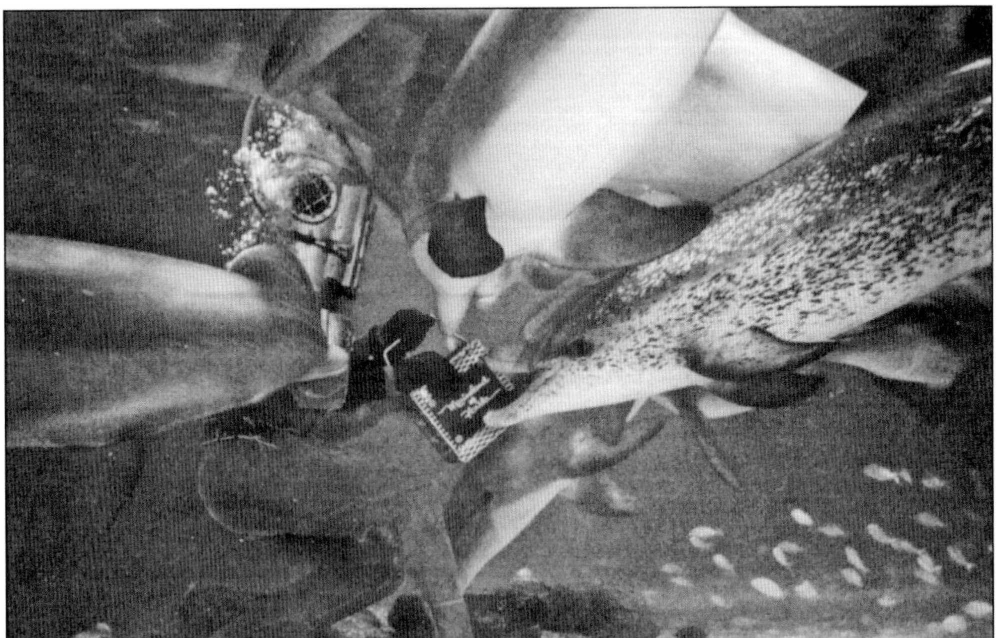

Three

Trendsetters

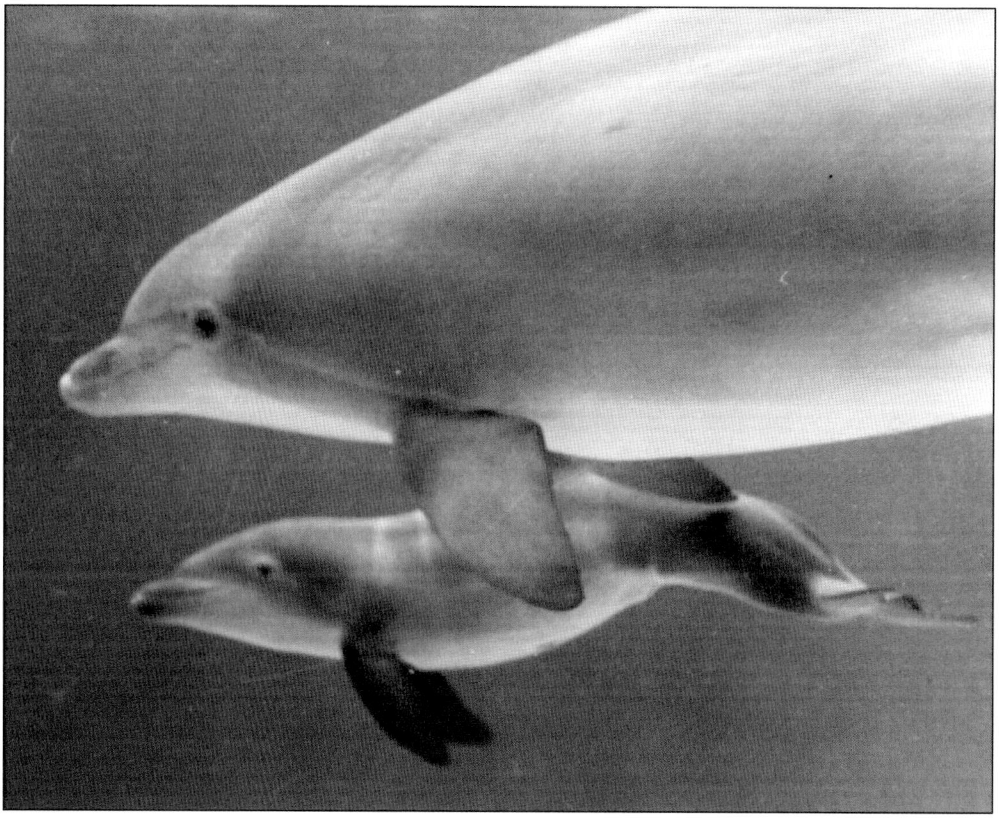

As the world's first oceanarium, Marine Studios was uniquely poised on the threshold of discovery. At the time, very little was known about the basic habits of most ocean animals, and as each new species arrived, it challenged the staff to uncover the unknown. This extraordinary opportunity to study marine life in close quarters was pivotal to increasing the world's understanding of the natural history and husbandry of life in the sea.

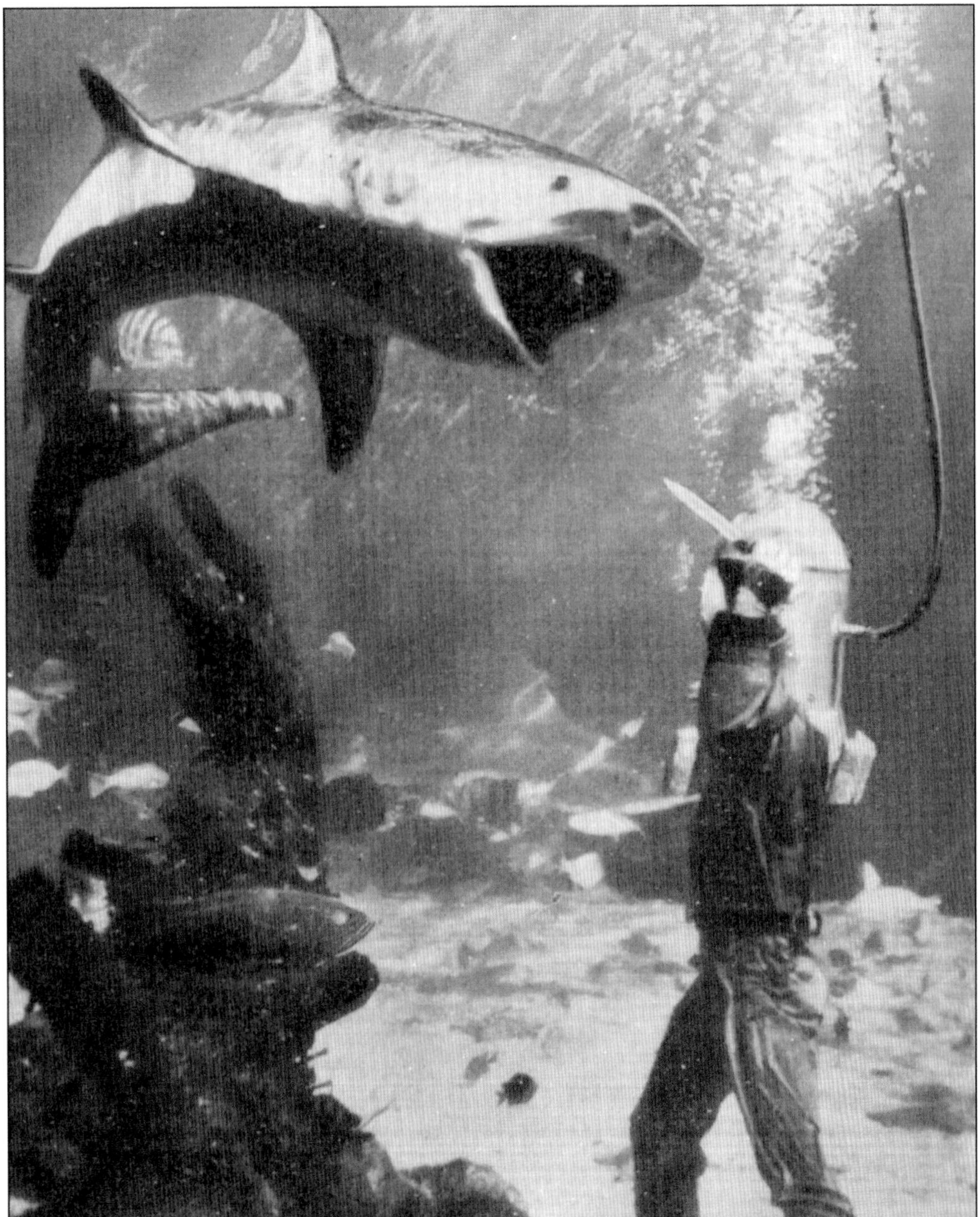

On February 22, 1938, the first shark arrived at Marine Studios and was placed in the oceanarium by motorist Robert Sabate, who used his car to deliver the six-and-a-half-foot, 150-pound ground shark from a beach seine net to the facility. She gave birth to nine pups several days later and lived for a total of nine days, setting a longevity record for the species. The previous record for the survival of a ground shark in an aquarium was only 36 hours. This event also marked the first time scientists had been able to observe the live birth of a shark under normal conditions. Eight months later, another record would be broken when the largest shark ever captured, an 11-and-a-half-foot-long ground shark, was introduced to the Marine Studios oceanarium. Ground sharks, also called whaler sharks, make up the largest scientific order of sharks, comprised of 270 different species, making it impossible to identify the exact species first introduced.

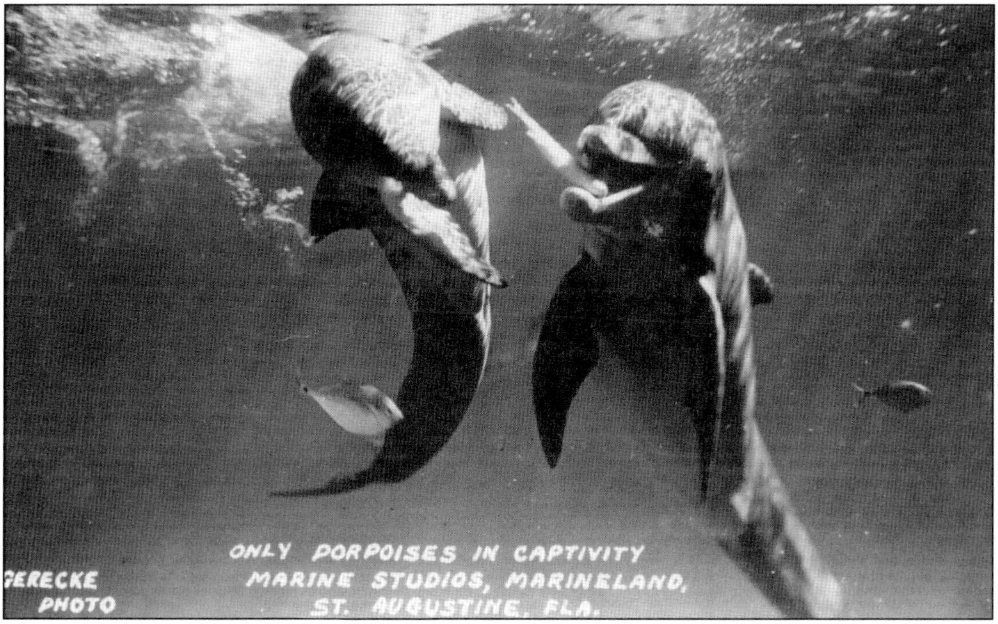

In anticipation of opening, new species were arriving on a daily basis. Two days after the first shark arrived, the first bottlenose dolphin was introduced to the circular oceanarium on February 24, 1938. A young male, he was collected in the San Sebastian River by Wallace Caswell. Joined shortly after by many others, they were the only bottlenose dolphins to be seen in any aquarium at the time.

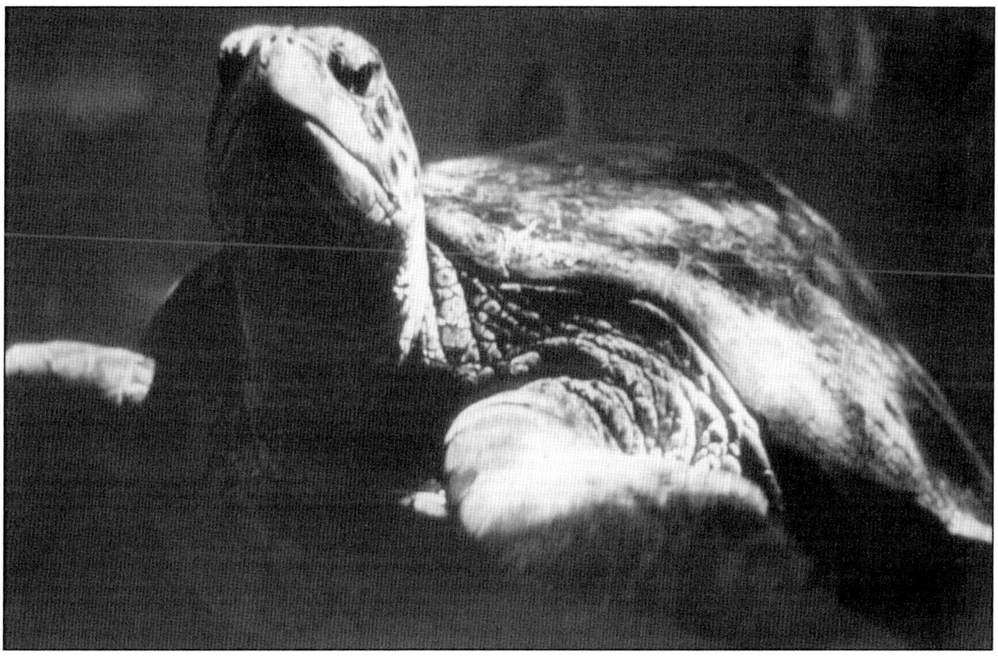

The next day, on February 25, 1938, two loggerhead sea turtles arrived. Collected in the Cayman Islands and transported via boat, train, and truck to Marine Studios, they were estimated to weigh 200 pounds each. As the story goes, the staff watched as the giant turtles swam by the porthole windows with a note affixed to one of their shells declaring, "Weight: 200 pounds, Price: $20."

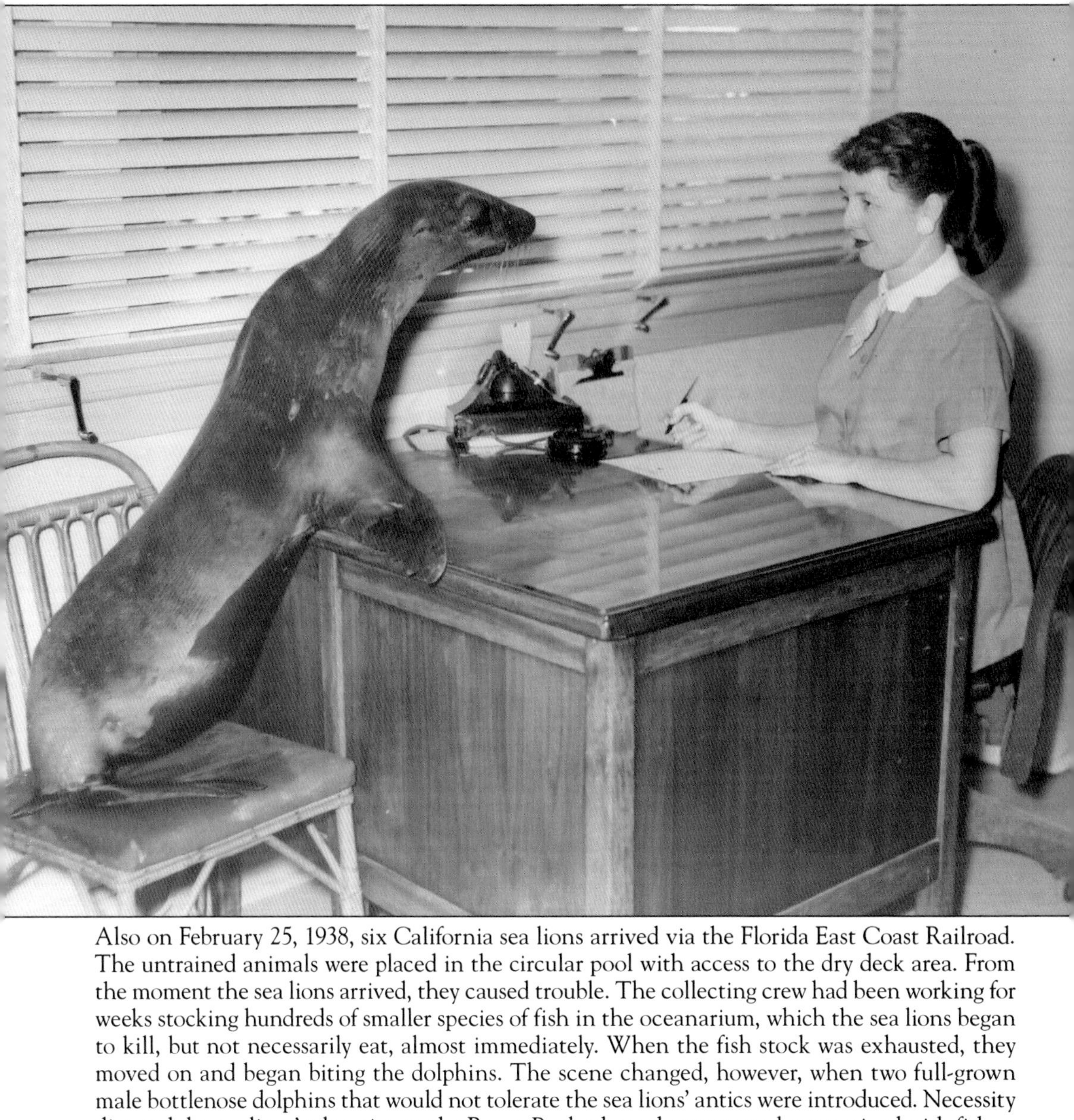

Also on February 25, 1938, six California sea lions arrived via the Florida East Coast Railroad. The untrained animals were placed in the circular pool with access to the dry deck area. From the moment the sea lions arrived, they caused trouble. The collecting crew had been working for weeks stocking hundreds of smaller species of fish in the oceanarium, which the sea lions began to kill, but not necessarily eat, almost immediately. When the fish stock was exhausted, they moved on and began biting the dolphins. The scene changed, however, when two full-grown male bottlenose dolphins that would not tolerate the sea lions' antics were introduced. Necessity dictated the sea lions' relocation to the Bayou Pool, where they were no longer mixed with fish or dolphins. The sea lions went on to win local fame as they frequently escaped the Bayou Pool and were found in neighboring waterways and marinas, not to mention the offices of pretty, young Marine Studios employees.

The animal collection expanded even more when the first penguins arrived—four African penguins from Robbin Island, South Africa, and one Magellanic penguin from the Straits of Magellan off the southern coast of Chile. As seen here, their plane was greeted by Arthur McBride and Ilia Tolstoy, who immediately got to work acclimating the new residents to their Marine Studios home.

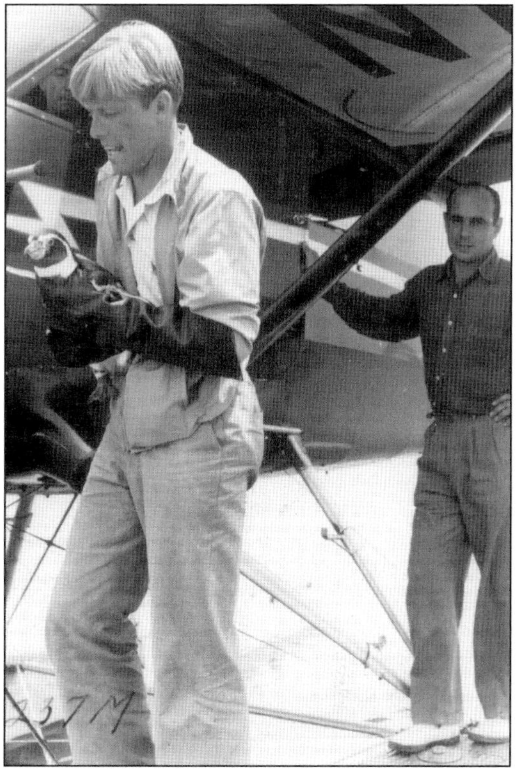

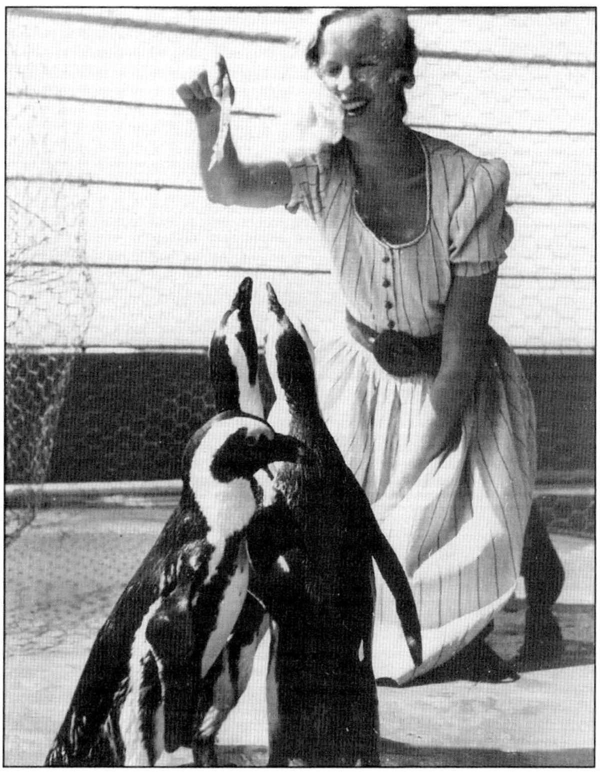

The penguins quickly became some of the favorite animals of the staff, guests, and visiting celebrities at Marine Studios because of their ease of care and comical nature. Their survival on the incredibly long journey is credited to the expert care of collector Ellis Jones, a friend of Tolstoy who, although retired, was the original animal collector for most zoos around the country.

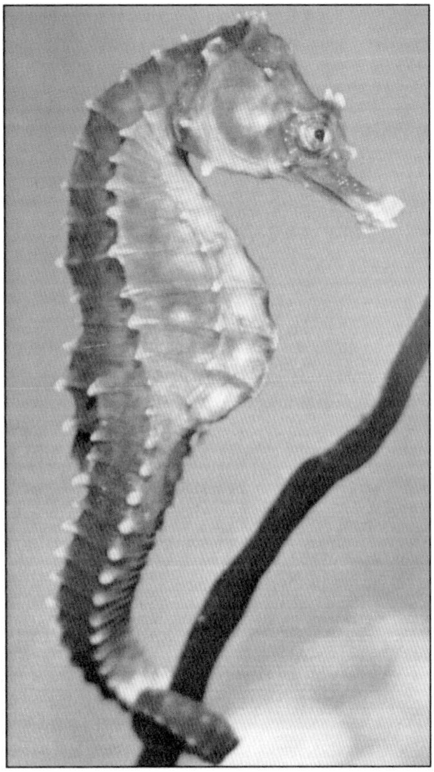

Many exciting events took place at Marine Studios, including the first stingray birth, resulting in an offspring named Martha. Seahorses also successfully gave birth at the oceanarium, producing several hundred babies during their first week in residence. But perhaps the most exciting day was when Marine Studios displayed a giant manta ray with a wingspan of over 13 feet. Placed in the circular oceanarium in 1940, it swam the perimeter of the pool 24 hours a day, disturbing the resting patterns of the dolphins and driving them to huddle in the center. Unable to tempt it to eat, the staff force-fed 35 fish to the manta ray, only to discover after its death that its stomach was full of tiny shrimp-like organisms called krill. Both in life and in death, there were scientific discoveries to be made.

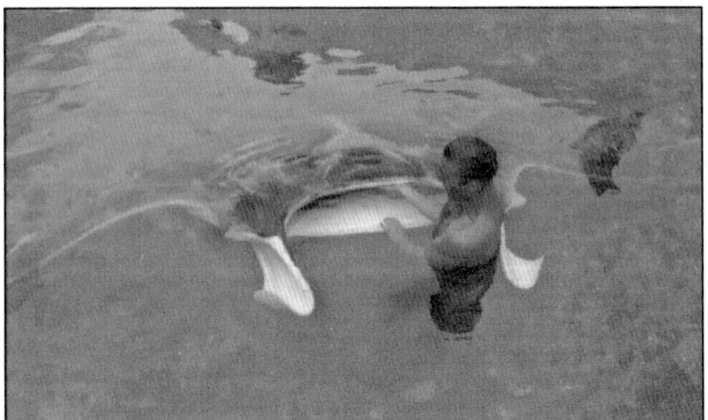

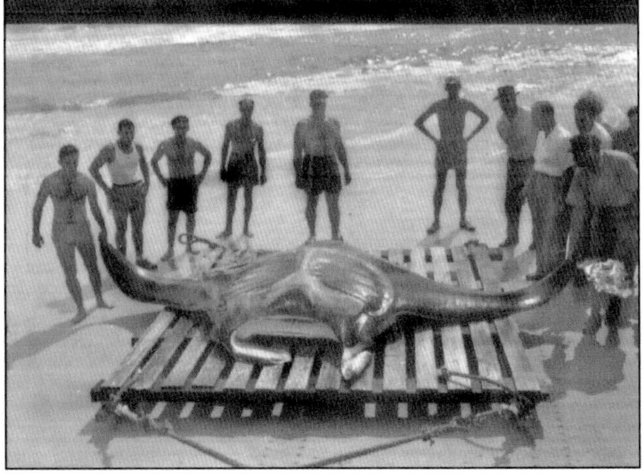

The first dolphin birth in the United States took place at Marine Studios on December 23, 1939. Although unsuccessful, it provided the staff with information that eventually led to the first successful dolphin birth in the world, which took place in the same pool on February 26, 1947. The female calf was named Spray. Her birth was a significant event. As Spray thrived under the care and nurturing of the Marine Studios staff, no one suspected she might set yet another record, but she did. On March 4, 1954, at seven years of age, Spray gave birth to Marine Studios' first grandbaby dolphin, a female calf named Peggy, after scientist and researcher Margaret "Peggy" Tavolga. Spray's mother, Mona, reportedly assisted with Peggy's rearing, marking three generations of dolphins in the Marine Studios oceanarium.

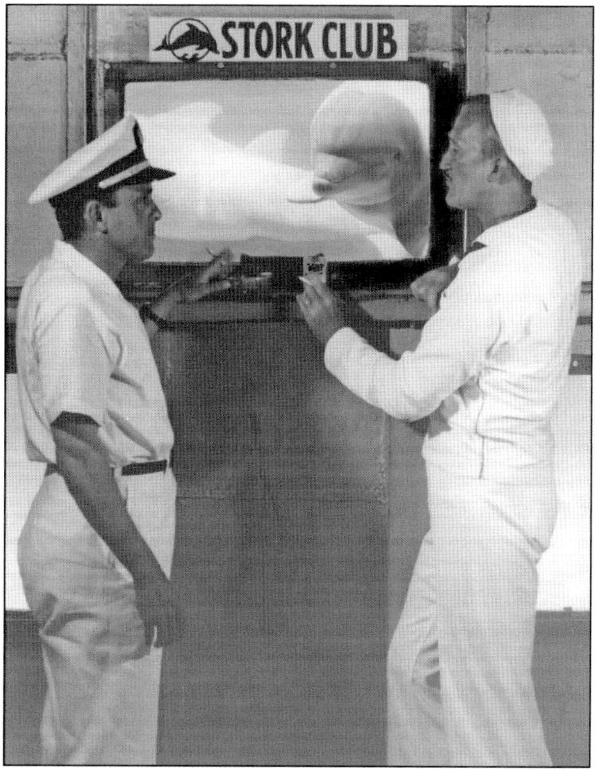

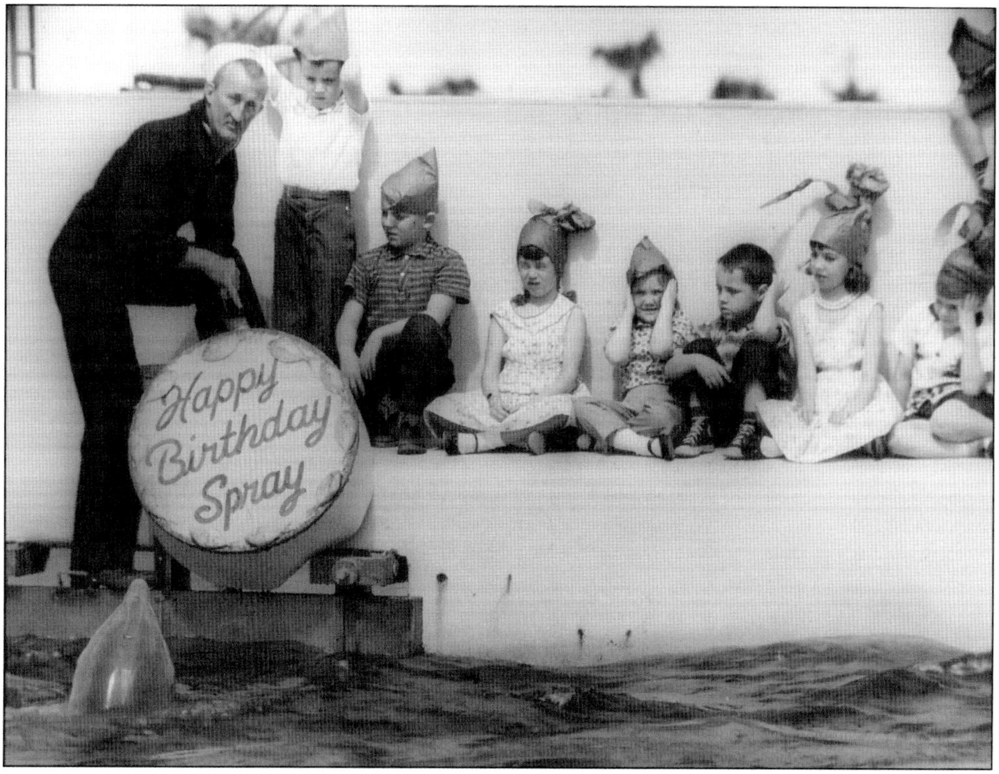

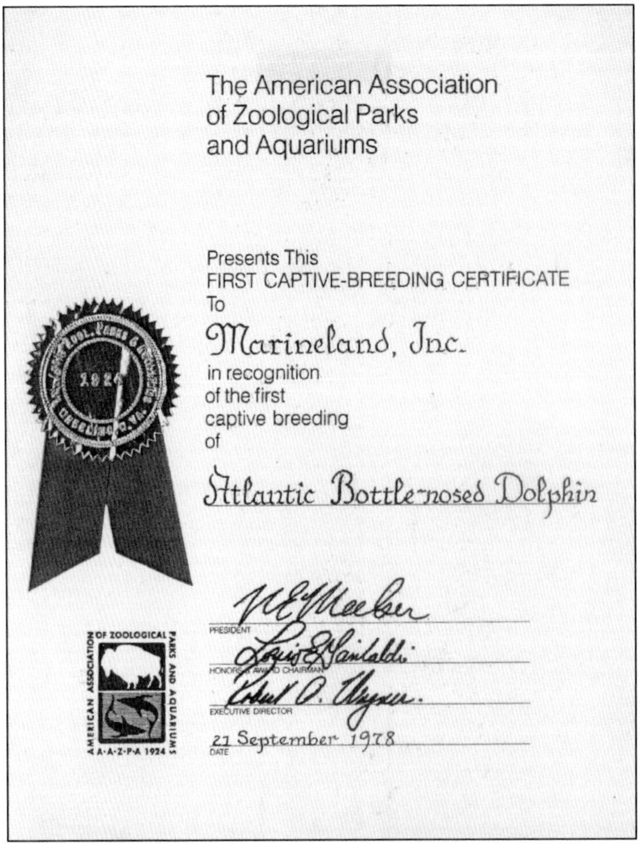

The American Association
of Zoological Parks
and Aquariums

Presents This
FIRST CAPTIVE-BREEDING CERTIFICATE
To

Marineland, Inc.

in recognition
of the first
captive breeding
of

Atlantic Bottle-nosed Dolphin

PRESIDENT

HONORS & AWARD CHAIRMAN

EXECUTIVE DIRECTOR

21 September 1978
DATE

AMERICAN ASSOCIATION OF ZOOLOGICAL PARKS AND AQUARIUMS
A·A·Z·P·A 1924

Marine Studios' success with the breeding and reproduction of Atlantic bottlenose dolphins earned acknowledgment and a certificate of recognition from the American Association of Zoological Parks and Aquariums in 1978. More important than the certificate were the discoveries made that would benefit zoos and aquariums around the world and the animals that lived there. Ages of reproductive maturity, length of gestation period, signs of labor, requirements for a healthy delivery, nutritional needs of a mother and calf, and calves' dependency on parental care were just some of the invaluable discoveries made at Marine Studios. By 1951, after only nine operational years, Marine Studios had proudly experienced 16 successful dolphin births.

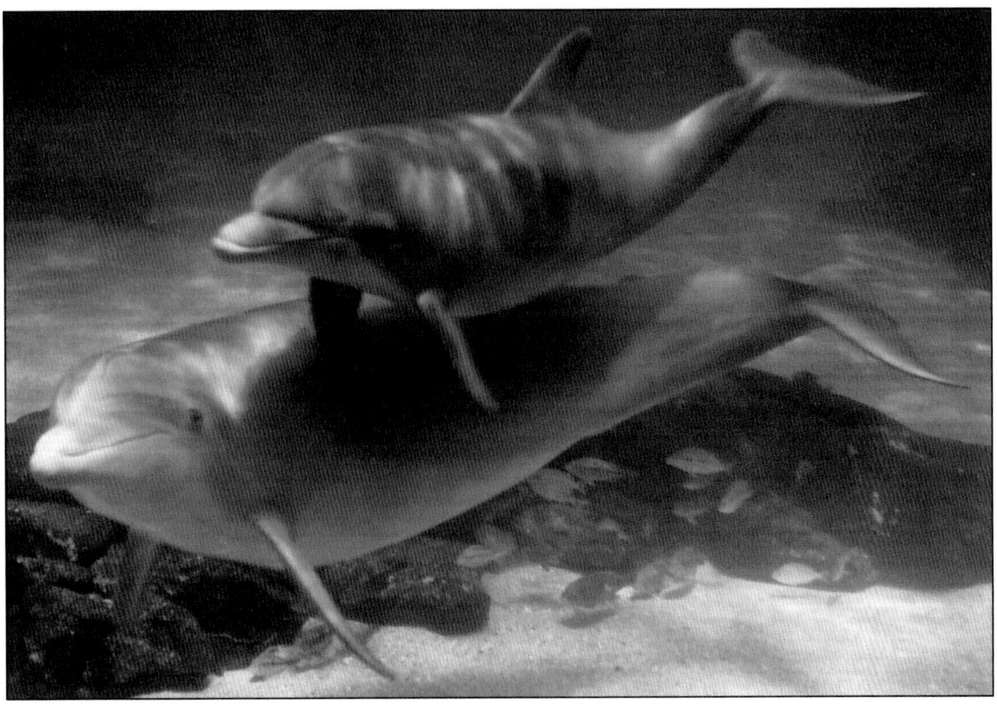

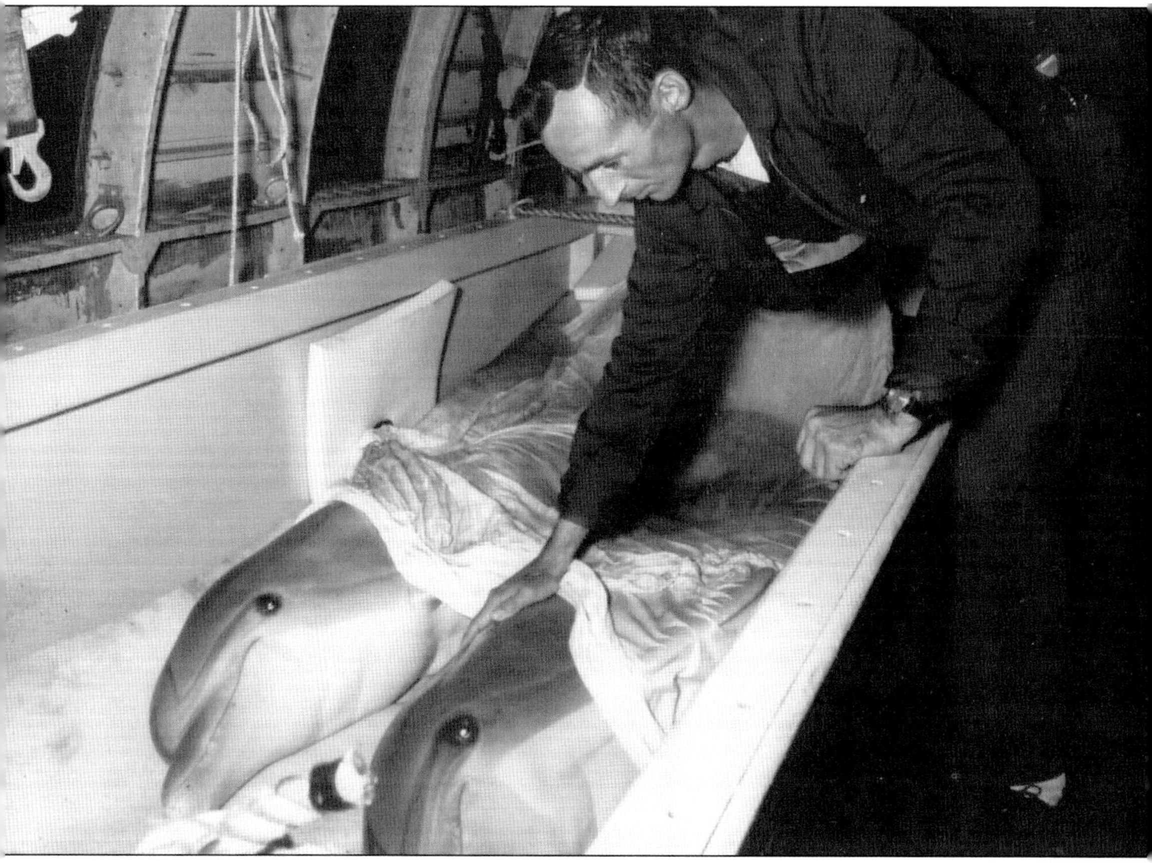

The first air transport of bottlenose dolphins took place on September 14, 1954, when four dolphins were transported on a Flying Tigers DC-3 airplane from Marine Studios to Marineland of the Pacific in California. Two of the dolphins, males Frank and Floyd, were born and raised at Marine Studios. The dolphins were trucked to the Daytona Beach airport, accompanied on their journey by curator F.G. Wood and general manager W.F. Rolleston. The dolphins were covered in wet burlap and traveled in a rubber-lined, cushioned box containing saltwater. The flight itself covered 2,200 miles and took 14 hours. It was suspected that because dolphins are deep divers, they might have difficulty with high altitudes, so the plane was flown as low as possible, while the temperature inside was kept at about 50 degrees. Another concern was the possibility of airsickness. In the end, it was agreed that the plane ride went smoothly, but it was the truck ride to Marineland of the Pacific that eventually made the dolphins sick.

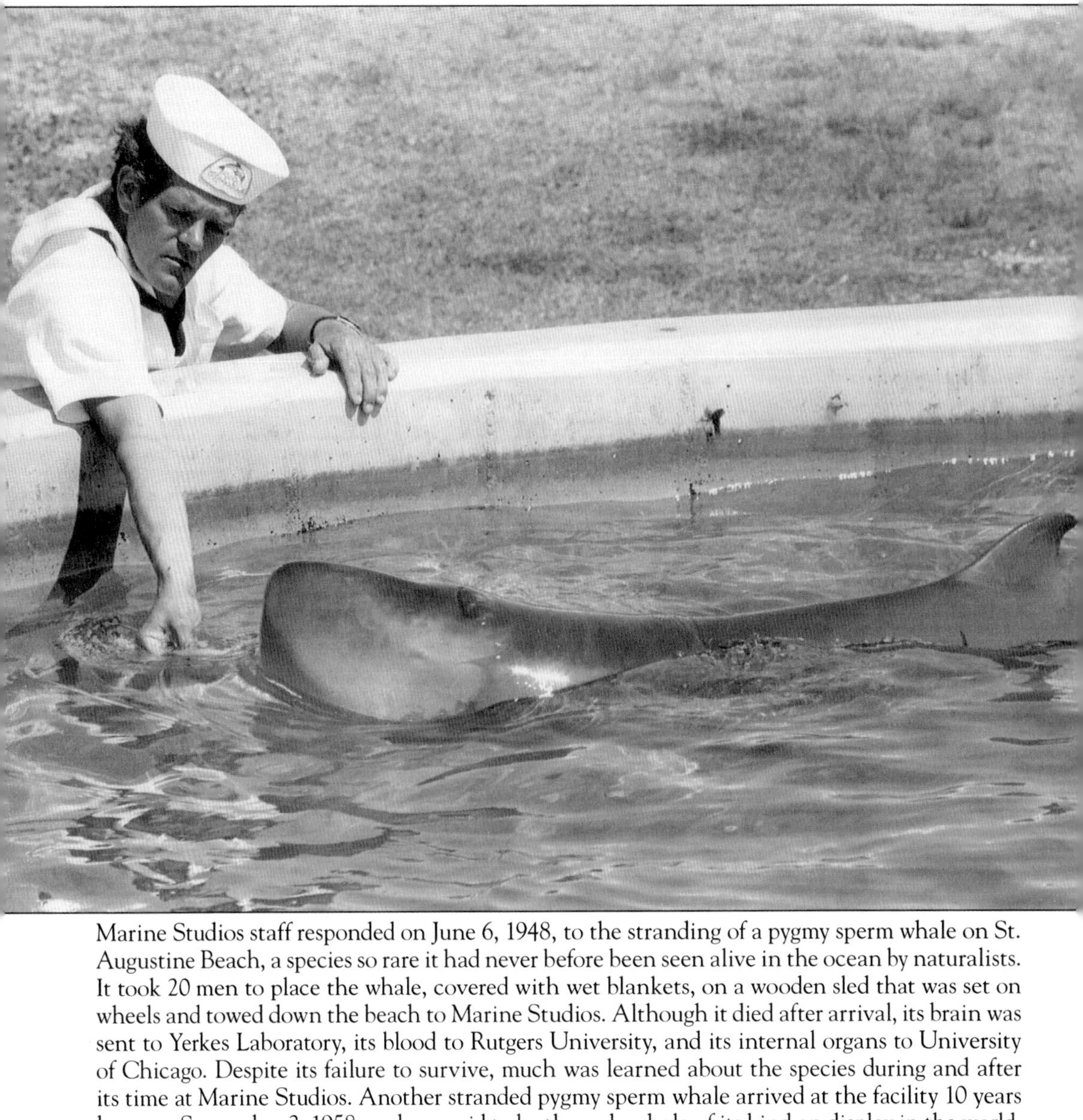

Marine Studios staff responded on June 6, 1948, to the stranding of a pygmy sperm whale on St. Augustine Beach, a species so rare it had never before been seen alive in the ocean by naturalists. It took 20 men to place the whale, covered with wet blankets, on a wooden sled that was set on wheels and towed down the beach to Marine Studios. Although it died after arrival, its brain was sent to Yerkes Laboratory, its blood to Rutgers University, and its internal organs to University of Chicago. Despite its failure to survive, much was learned about the species during and after its time at Marine Studios. Another stranded pygmy sperm whale arrived at the facility 10 years later, on September 2, 1958, and was said to be the only whale of its kind on display in the world. It is shown here with Tommy DeVoe. Marine Studios provided a monetary reward to the four youths vacationing from Georgia who saved the whale by reporting its stranding.

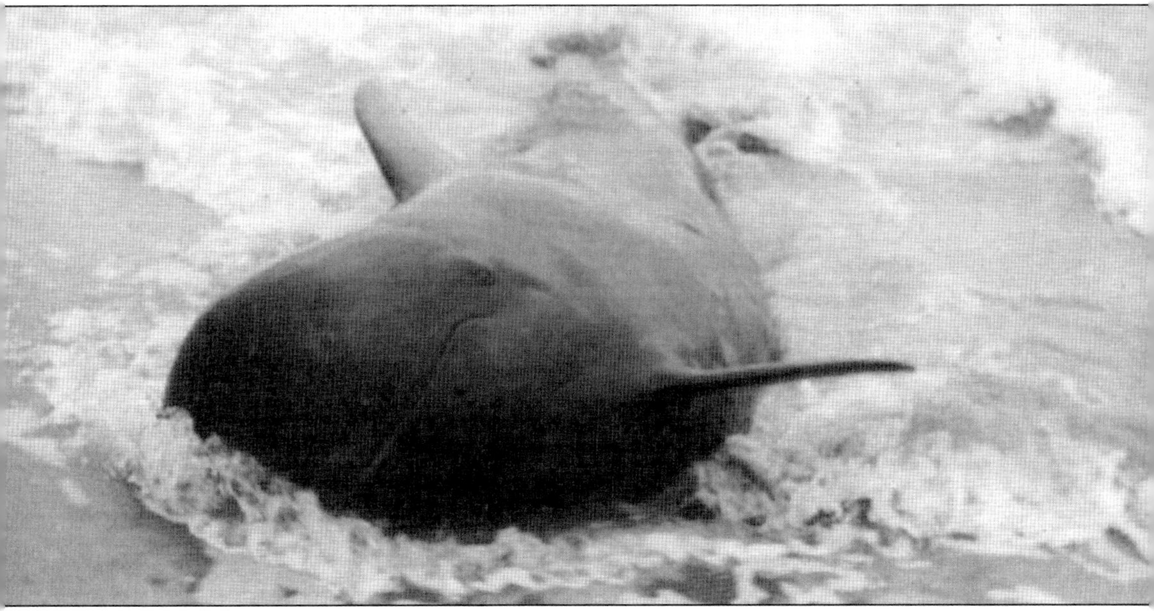

Marine Studios welcomed its first pilot whale on October 5, 1948. Following a mass stranding of 44 pilot whales on Crescent Beach, the staff chose four of the strongest animals for transport to the oceanarium, but only one survived. Introduced to the circular pool with the dolphins, the young male pilot whale was named Herman. Uncertain what to feed a pilot whale, the staff offered him fish and blue crabs, which he refused; however, the squid they offered was readily accepted. In fact, Herman would eat up to 30 pounds of squid each day. He was a favorite at the oceanarium and was usually accompanied by bottlenose dolphins Mona and Spray. Herman taught the staff that pilot whales were nocturnal and that his most successful feedings took place in the early-morning hours. Later, curator Arthur McBride reflected that he should have deduced that pilot whales were nocturnal by the dark color of their skin, which could have led to the discovery that they fed on squid, as they rise to the ocean surface each evening.

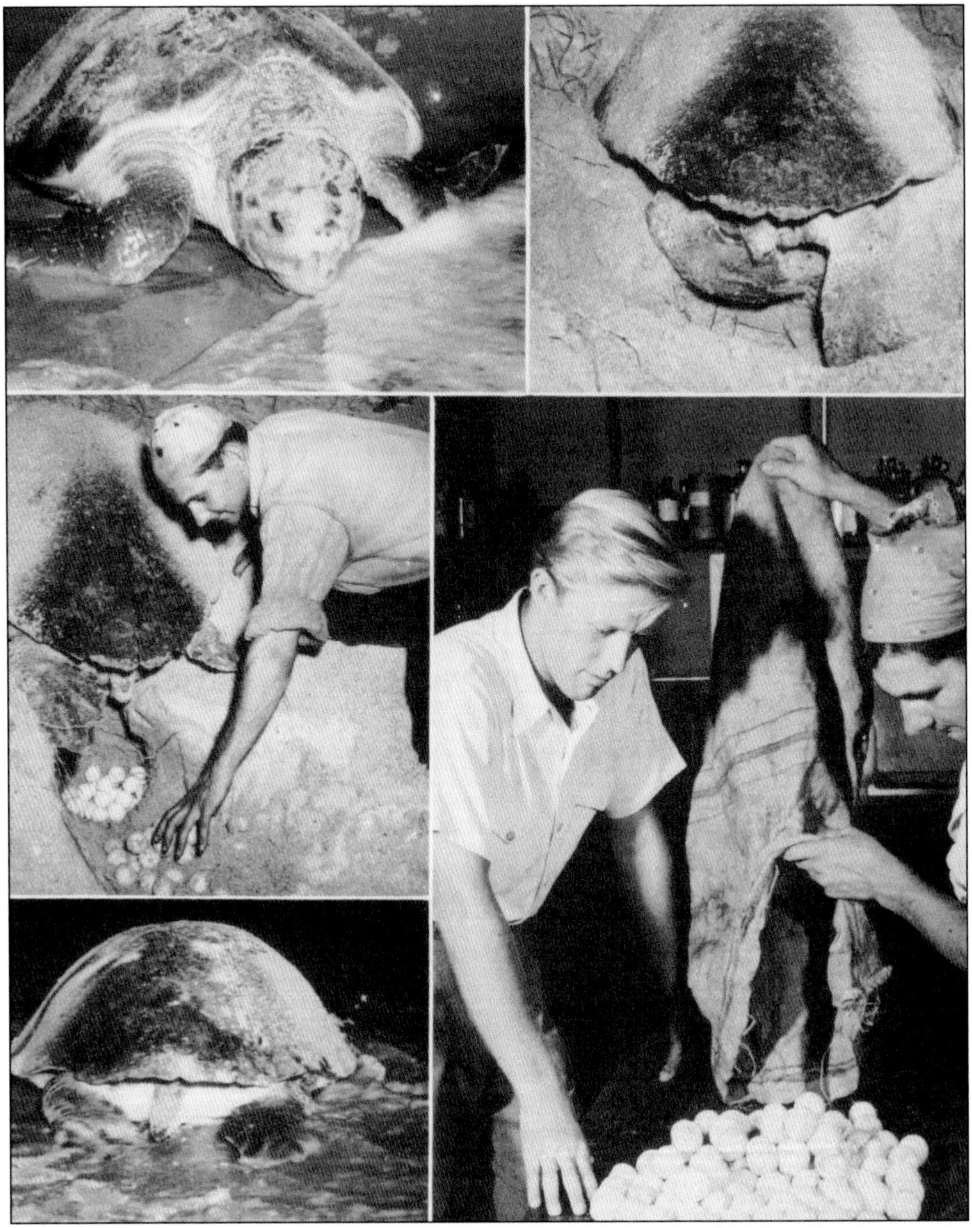

The rehabilitation of marine mammals was only part of Marine Studios' broader conservation focus. Appalled by the practice of sea turtle killings on local beaches, Marine Studios joined forces with the Florida State Department of Conservation in 1940 to offer a cash reward of $100 leading to the arrest and conviction of persons guilty of slaughtering sea turtles laying eggs on the beaches of St. Johns, Flagler, and Volusia Counties. Impatient egg hunters were cutting the eggs directly from the females, killing them in their eagerness to obtain the eggs. Marine Studios took a strong stance, calling this practice "inhumane viciousness that must be stopped," and voiced their concern that the ruthless slaughter of these creatures would inevitably end in the extinction of the species. Years later, when several hundred loggerhead turtle eggs were found on the beach at Marineland, they were taken to Marine Studios, where 108 baby turtles were successfully hatched after a 54-day incubation period.

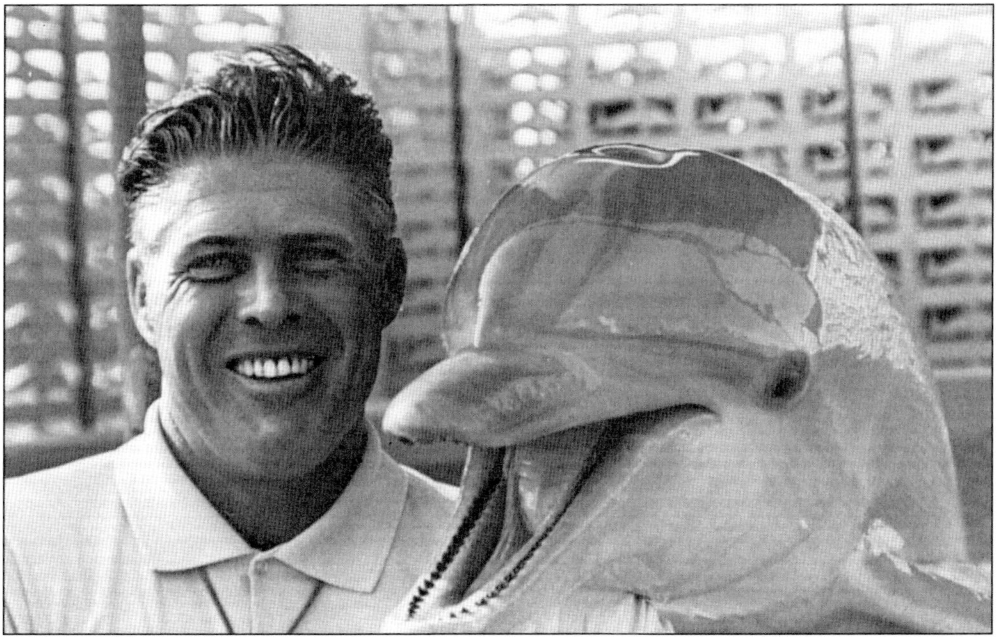

In 1964, the Marine Studios collection boat stumbled upon a most unique and unusual sighting, a bottlenose dolphin with a very light phase of skin pigmentation, which they dubbed "champagne" or "blonde." Between the years of 1964 and 1968, three blonde dolphins called Marine Studios home: female Lilly Champagne (shown above with trainer Fred Lyons) and males Big Boy and Rusty. The most beloved of the blonde dolphins was Lilly (below right), who formed a long-lasting bond with bottlenose dolphin Nellie (below left), a friendship that endured for 45 years. Although Lilly left no progeny, blonde dolphin Rusty did breed with Nellie, which resulted in the birth of a normal-colored male dolphin named Sunny. Although Sunny himself is not blonde, he does continue to carry the blonde gene today.

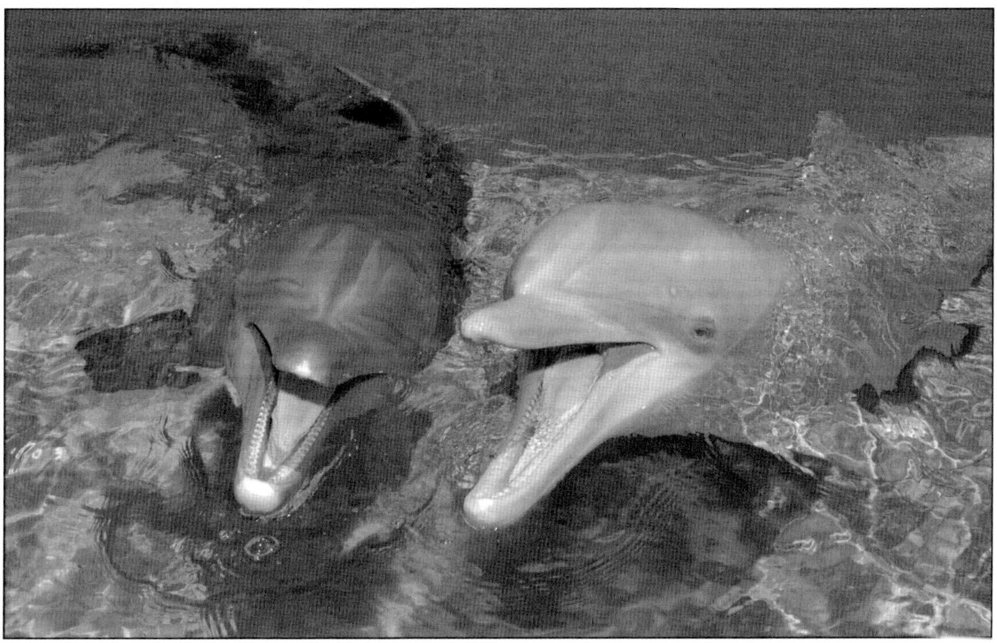

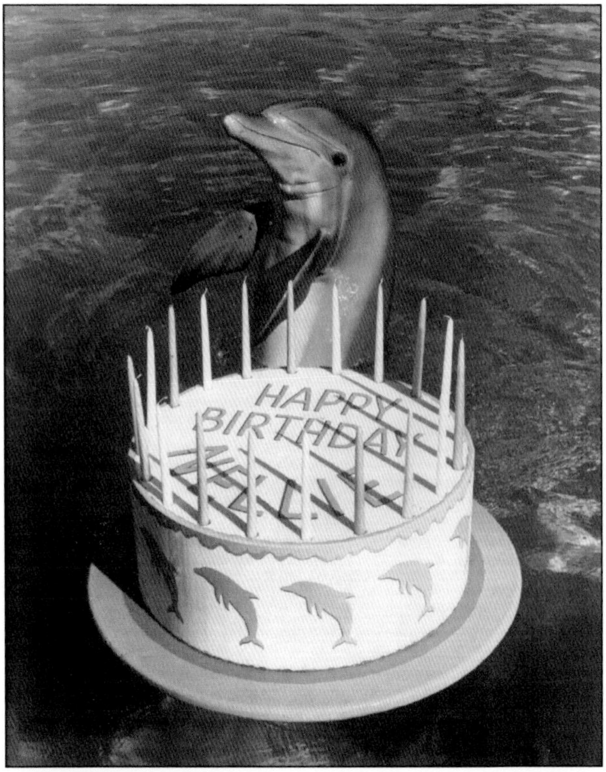

By far, the most remarkable and beloved bottlenose dolphin at Marine Studios is Nellie. Born in the circular oceanarium on February 27, 1953, to mother Susie and father Happy, Nellie holds the distinction of being the oldest and longest-living bottlenose dolphin in any oceanarium in the world. Nellie was named after Nellie Mae Neighbors, who was an employee at the park for over 10 years.

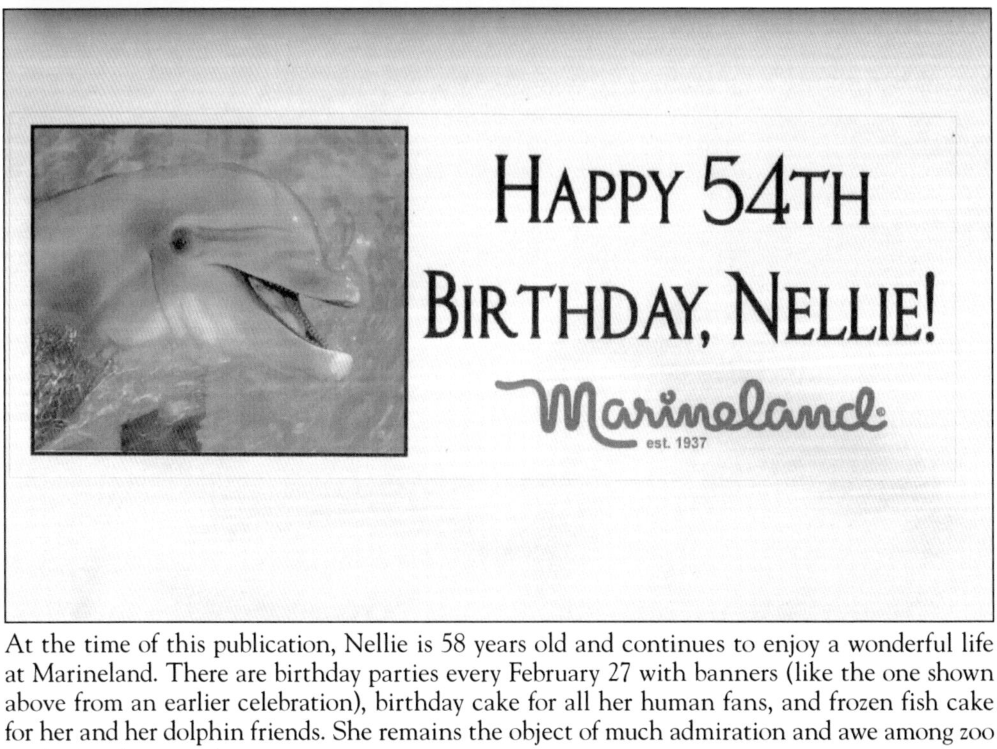

At the time of this publication, Nellie is 58 years old and continues to enjoy a wonderful life at Marineland. There are birthday parties every February 27 with banners (like the one shown above from an earlier celebration), birthday cake for all her human fans, and frozen fish cake for her and her dolphin friends. She remains the object of much admiration and awe among zoo professionals everywhere.

Four

RESEARCHERS

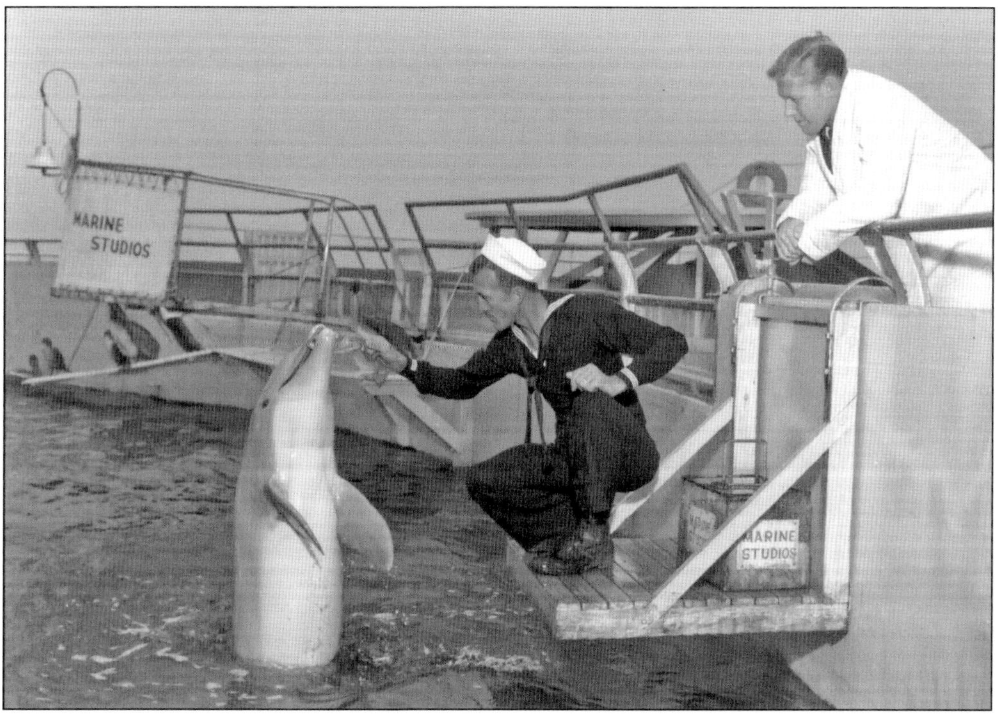

In the spirit of discovery that fueled Marine Studios from the beginning, research and science were always at the forefront of daily activities. Beginning with the first curator, Arthur McBride (right), and continuing to this day with both resident and visiting scholars, the oceanarium has hosted some of the great minds in marine research whose discoveries have led to advances in conservation and improvements in animal care.

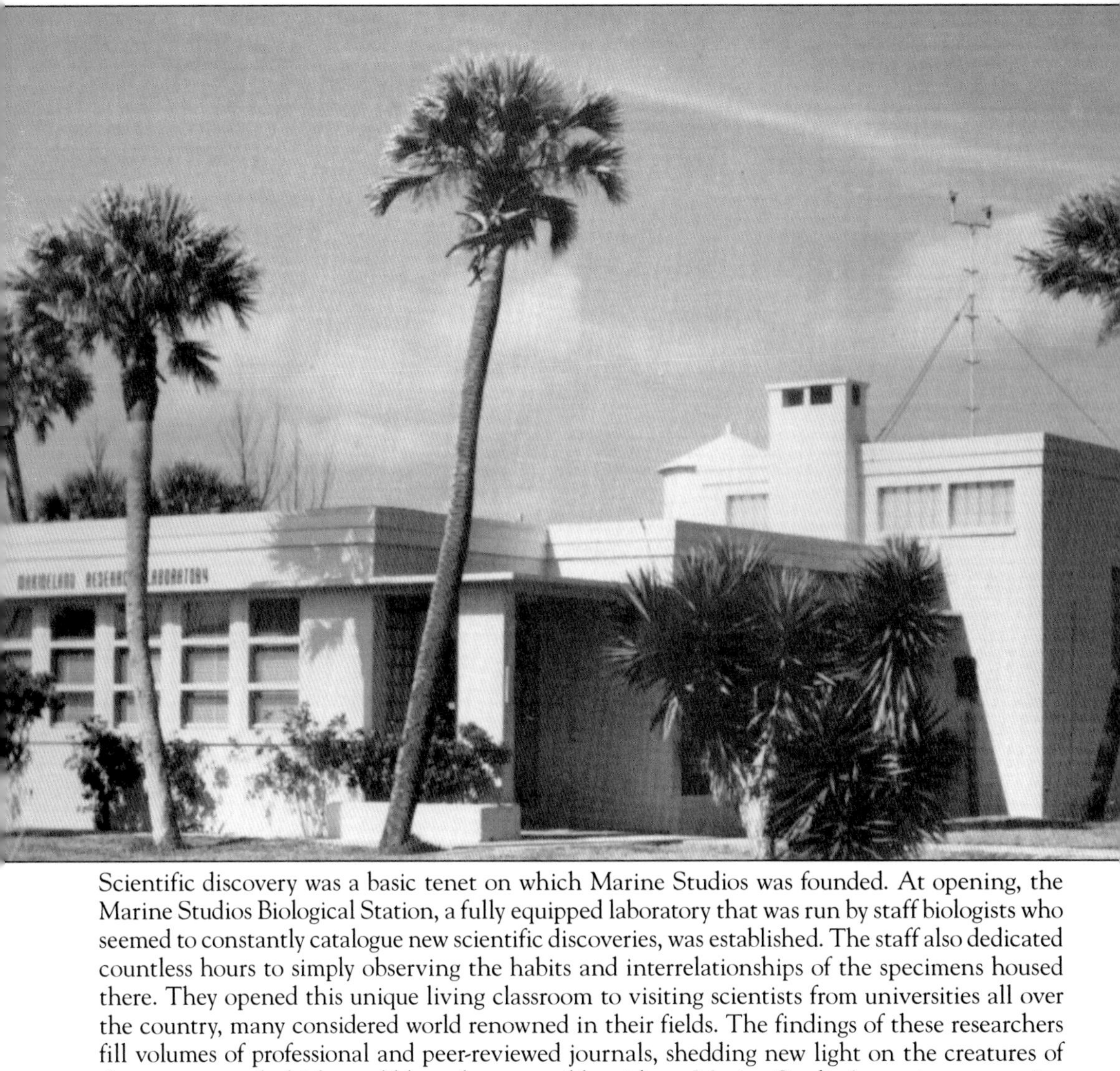

Scientific discovery was a basic tenet on which Marine Studios was founded. At opening, the Marine Studios Biological Station, a fully equipped laboratory that was run by staff biologists who seemed to constantly catalogue new scientific discoveries, was established. The staff also dedicated countless hours to simply observing the habits and interrelationships of the specimens housed there. They opened this unique living classroom to visiting scientists from universities all over the country, many considered world renowned in their fields. The findings of these researchers fill volumes of professional and peer-reviewed journals, shedding new light on the creatures of the sea, none of which would have been possible without Marine Studios' amazing generosity. A new laboratory, aptly named the Marineland Research Laboratory, was built in 1952 as the science program grew and expanded. It was well equipped, containing a library, three laboratories, a chemical room, a photographic darkroom, and a large aquarium room for the preservation of live specimens. In true dedication to education and learning, visiting scientists and university students were invited to use the new laboratory free of charge.

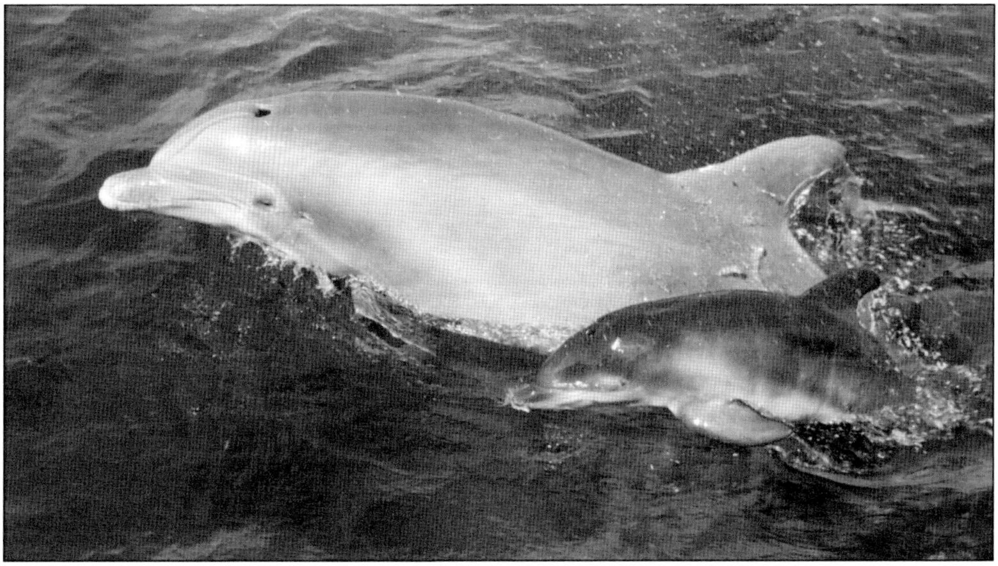

Marine Studios' first curator, Arthur McBride, was a passionate scientist whose interests spanned many species and topics. Originally from New York, he and his wife, Betty, lived on property at Marine Studios. He received his degree from the University of Florida in 1945. Some of his best-known work involved studies of bottlenose dolphin behavior, husbandry, and reproduction. In 1951, he coauthored a landmark paper with researcher Henry Kritzler titled *Observations on Pregnancy, Parturition, and Post-Natal Behavior in the Bottlenose Dolphin*. The paper outlined Marine Studios' experiences in caring for the very first bottlenose dolphin pregnancies in a zoological setting and included the rearing of five healthy dolphin calves, but was published posthumously. In 1949, McBride was diagnosed with Bright's disease and tragically died shortly thereafter at only 34 years of age.

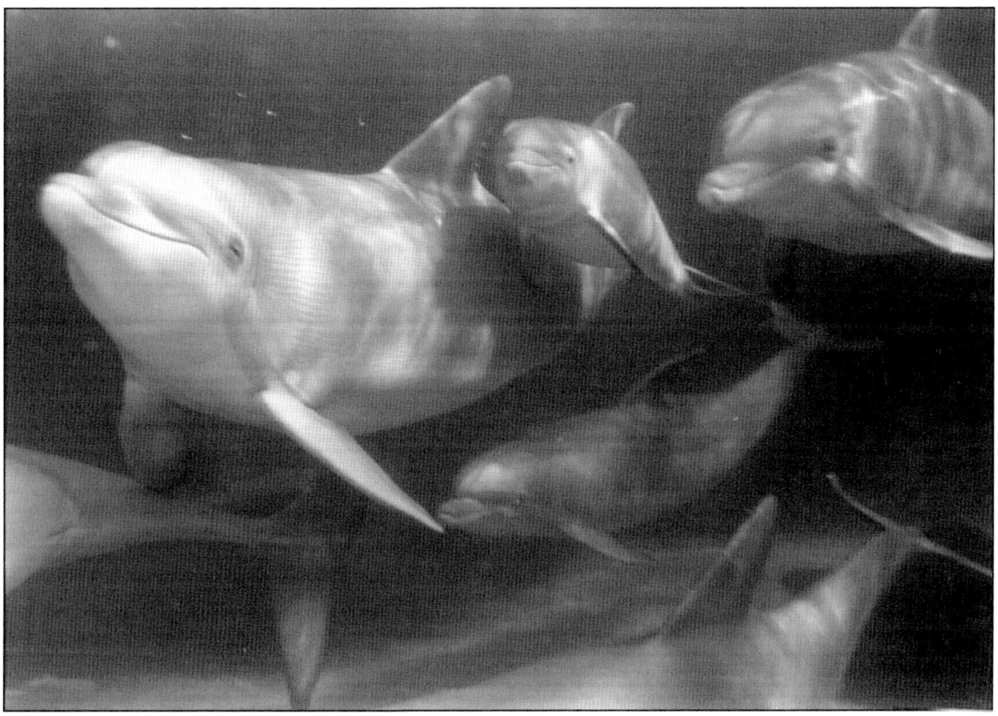

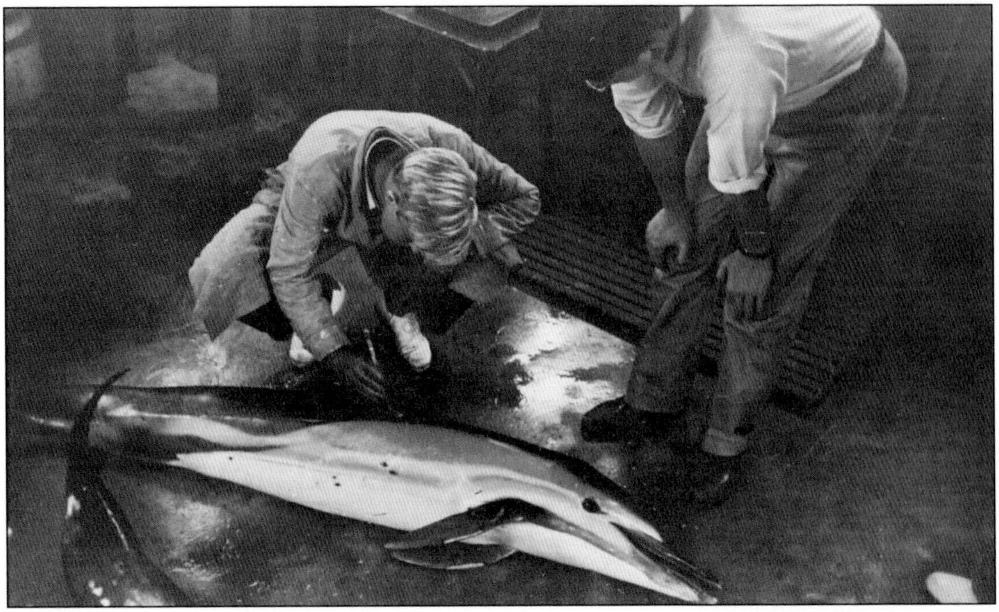

Eager to share his discoveries with more than just his peers, Arthur McBride's findings were published in popular magazines to educate the public about the fascinating life of the bottlenose dolphin. With this goal in mind, "Meet Mister Porpoise" came out in *Natural History* magazine in 1940. Some of McBride's findings concerning dolphin reproduction and breeding were considered so controversial in 1940s America that potential publishers petitioned to omit the information. Purity of science being McBride's hallmark, censorship was not to be tolerated. He turned away from popular magazines and back to the academic journals. In 1948, he coauthored a paper with colleague D.O. Hebb titled "Behavior of the Captive Bottle-nose Dolphin" in the *Journal of Comparative and Physiological Psychology*. The paper outlined many new discoveries about bottlenose dolphin communication, intelligence, reproduction, and social hierarchies.

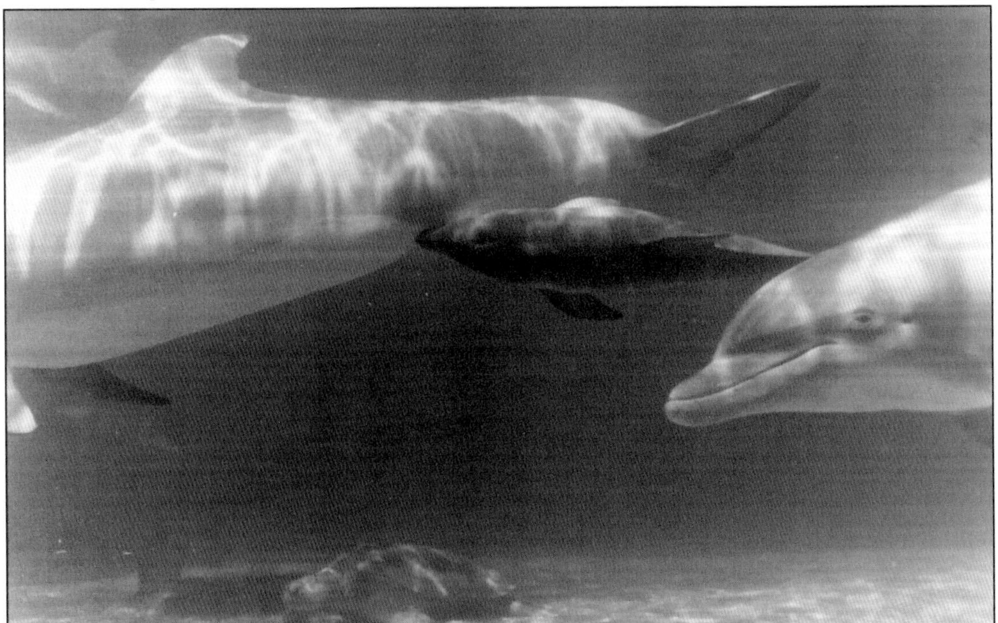

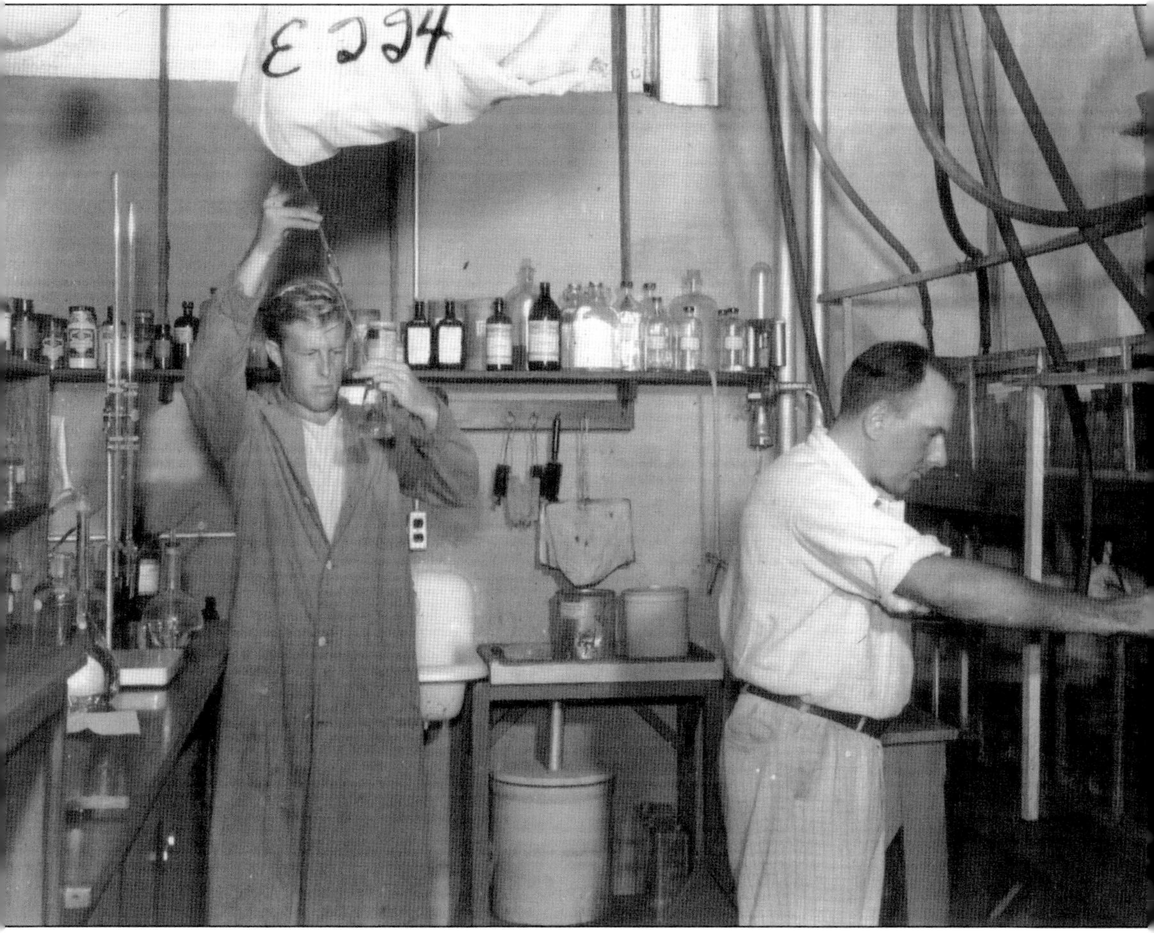

Biologists Dr. P.F. Scholander and Dr. S.W. Grinnell from the Department of Biology at Swarthmore College in Pennsylvania joined colleague Dr. Laurence Irving of the Marine Biological Station in Marineland to study the way bottlenose dolphins' breathe at Marine Studios. An electrocardiograph machine for measuring heartbeats was used for the first time on a dolphin, the sensors of which were fastened to the animals with the aid of women's brassieres. The scientists hoped to discover what happened when a dolphin held its breath while diving and to understand how they were able to hold their breath for such lengths of time. Tests run on seven newly collected dolphins showed normal heart rates of 80 beats per minute, which decreased to 40 beats per minute during submersion and to 30 beats per minute on deeper dives. Dr. Irving would later report these results to the National Academy of Sciences and publish the findings in 1941 as "The Respiration of the Porpoise" in the *Journal of Cellular and Comparative Physiology*.

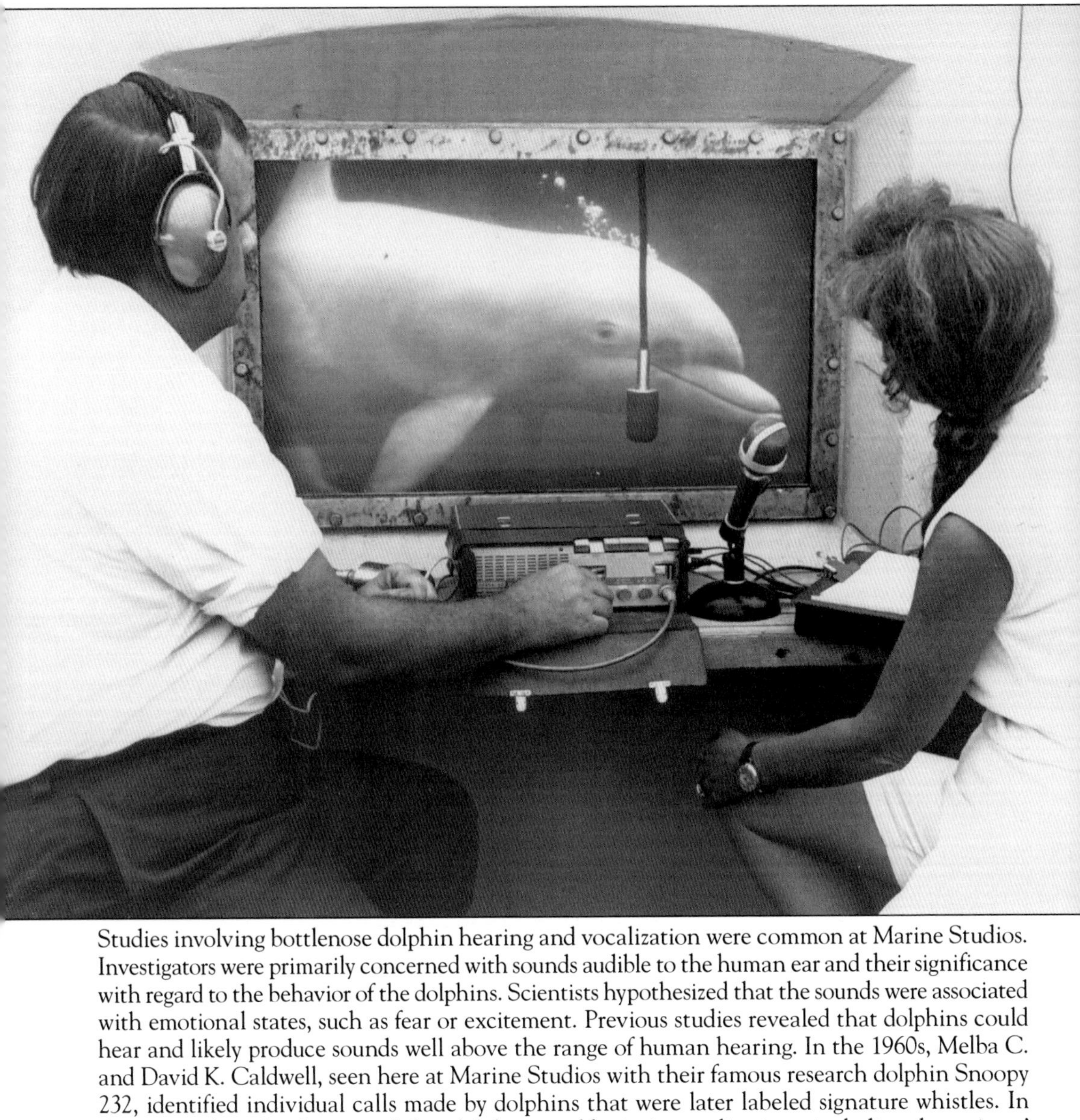

Studies involving bottlenose dolphin hearing and vocalization were common at Marine Studios. Investigators were primarily concerned with sounds audible to the human ear and their significance with regard to the behavior of the dolphins. Scientists hypothesized that the sounds were associated with emotional states, such as fear or excitement. Previous studies revealed that dolphins could hear and likely produce sounds well above the range of human hearing. In the 1960s, Melba C. and David K. Caldwell, seen here at Marine Studios with their famous research dolphin Snoopy 232, identified individual calls made by dolphins that were later labeled signature whistles. In the 1970s, they also discovered that dolphins could mimic vocalizations, including the trainers' whistles. The studies focused primarily on the dolphins' ability to produce sounds underwater. There were many other visiting scientists who studied dolphin intelligence and sound production at Marine Studios, including such notables as William E. Schevill, Winthrop N. Kellogg, and John C. Lily.

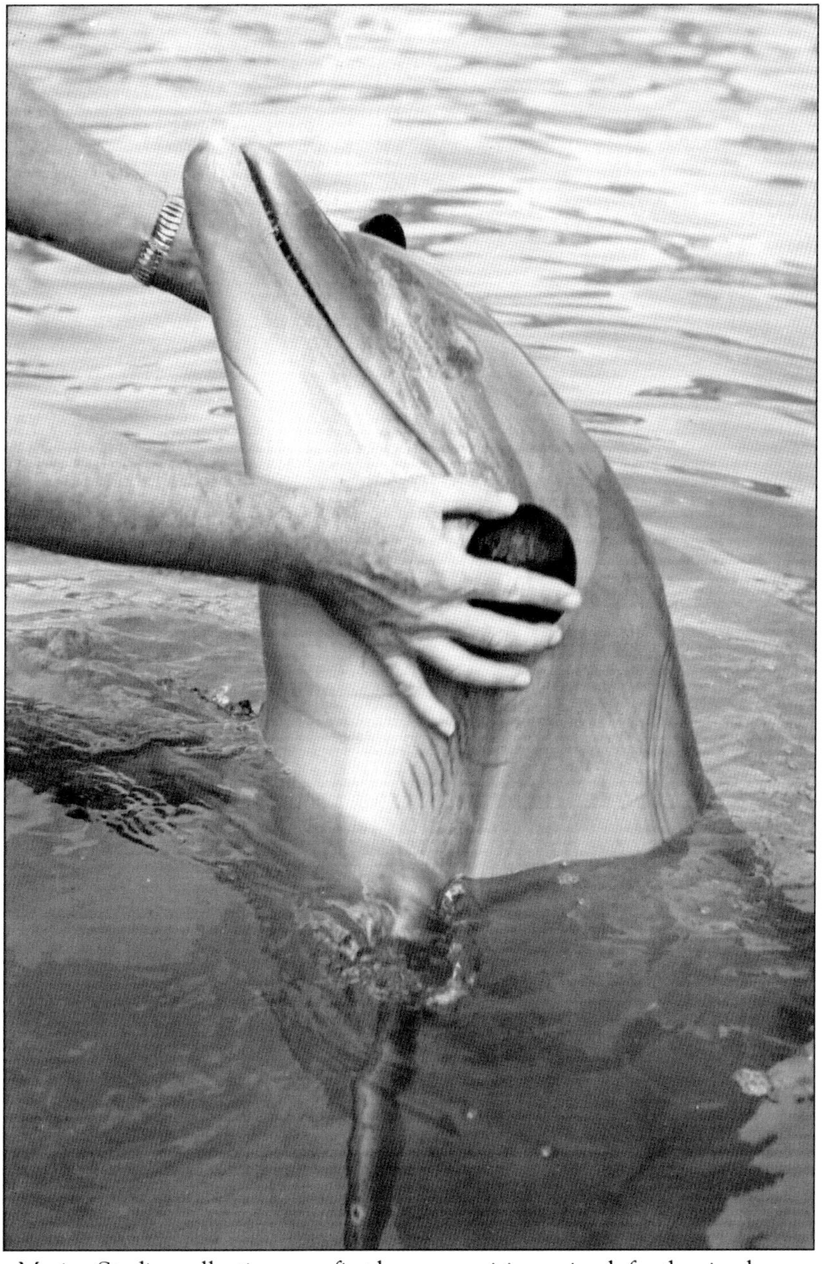

When the Marine Studios collecting crew first began acquiring animals for the circular oceanarium, a strange thing happened. The dolphins were evidently able to detect nets at some distance, even in murky water and on dark nights. This was hard for the collectors to understand. Due to poor visibility in the water, the collectors assumed the dolphins could not see the nets, yet the dolphins appeared to know that they were there. This led Arthur McBride to speculate as early as 1947 that dolphins used echolocation to navigate. Research to reproduce these blind settings was achieved with the use of soft rubber eyecups, like those in the photograph above. With the dolphins' eyes covered, researchers at Marine Studios tested and studied their ability to navigate using echolocation.

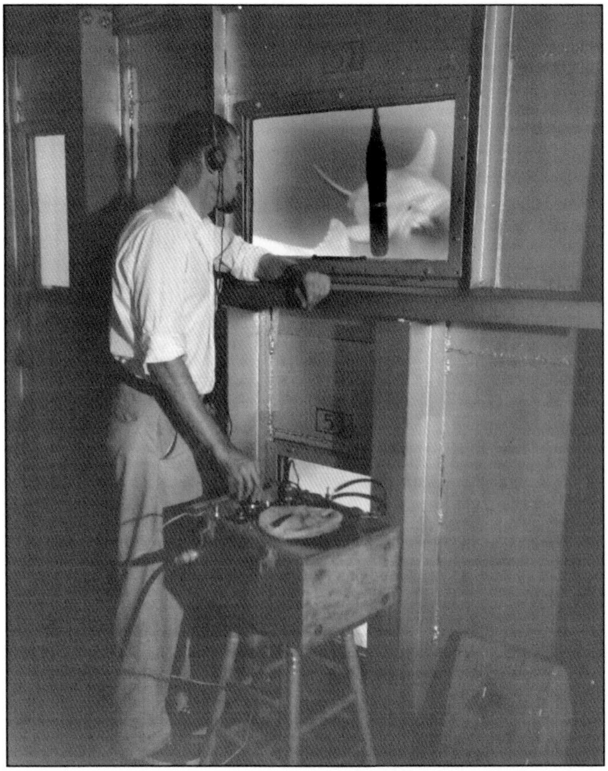

Early researchers at Marine Studios were greatly impressed with this remarkable ability of the bottlenose dolphin to seemingly navigate blindly. The observations that had been made during collecting seemed to provide good evidence that dolphins, like bats, used reflected sounds to detect obstacles, but much research was needed before a definitive answer could be reached.

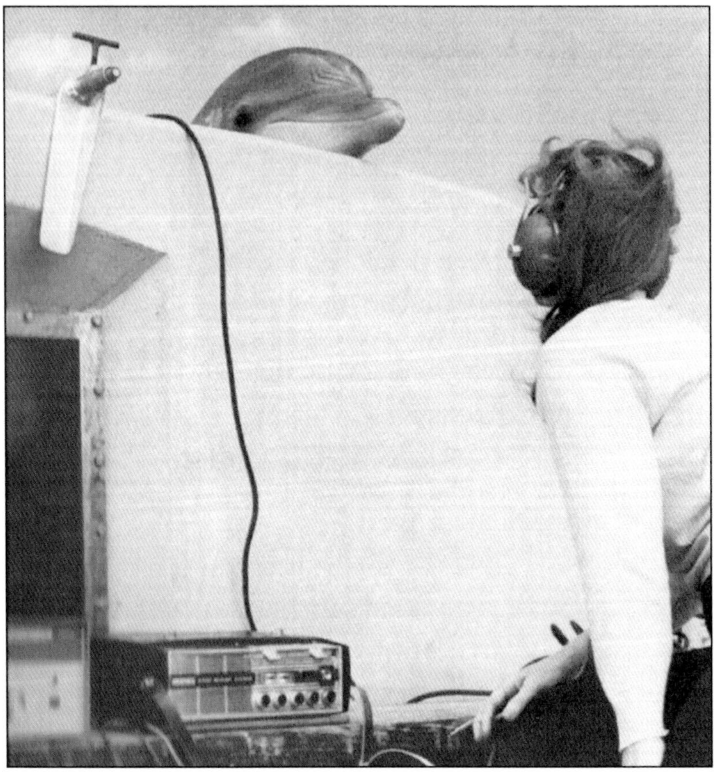

When a simple underwater microphone, known as a hydrophone, was placed in the water, researchers discovered many new and unusual dolphin sounds. The lower decks of the oceanarium were said to be the perfect place for listening to the dolphins' click and whistle sounds, which could often be heard reverberating through the pool wall and into the hallways during the quiet morning or evening hours.

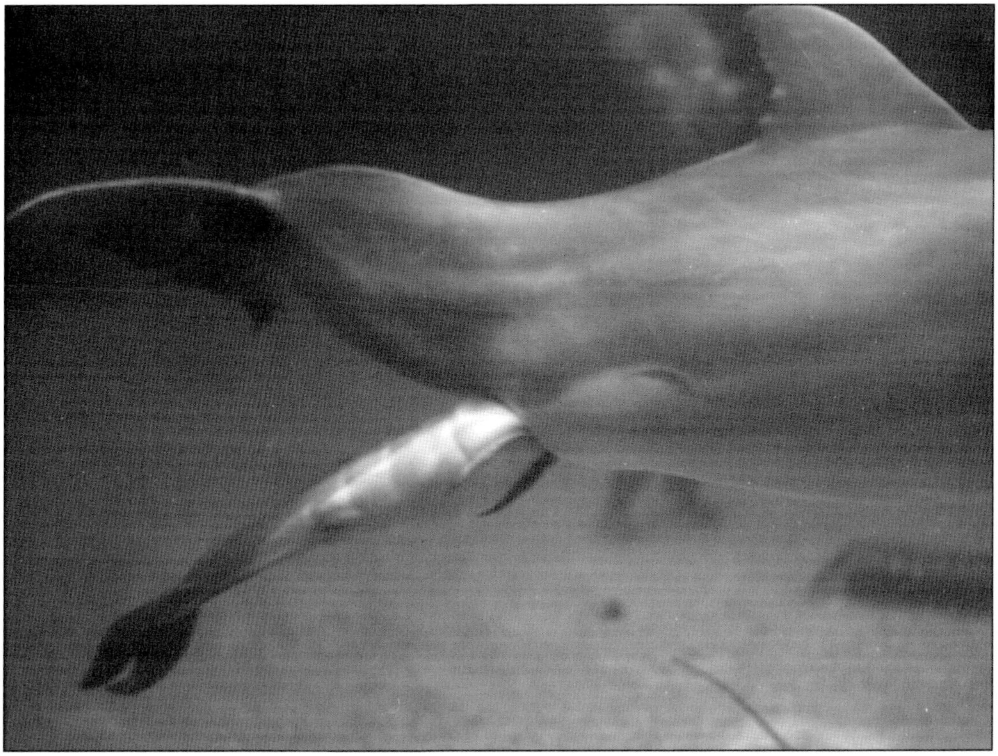

Besides sound production, Marine Studios' bottlenose dolphins were famous for their successful reproduction. Biologists Margaret "Peggy" Tavolga and Frank S. Essapian conducted extensive studies in this area. In 1953, Essapian published his work in *Natural History* magazine, in an article titled "The Birth and Growth of a Porpoise." In honor of Tavolga's contributions, the first grandbaby dolphin calf born at Marine Studios was named Peggy.

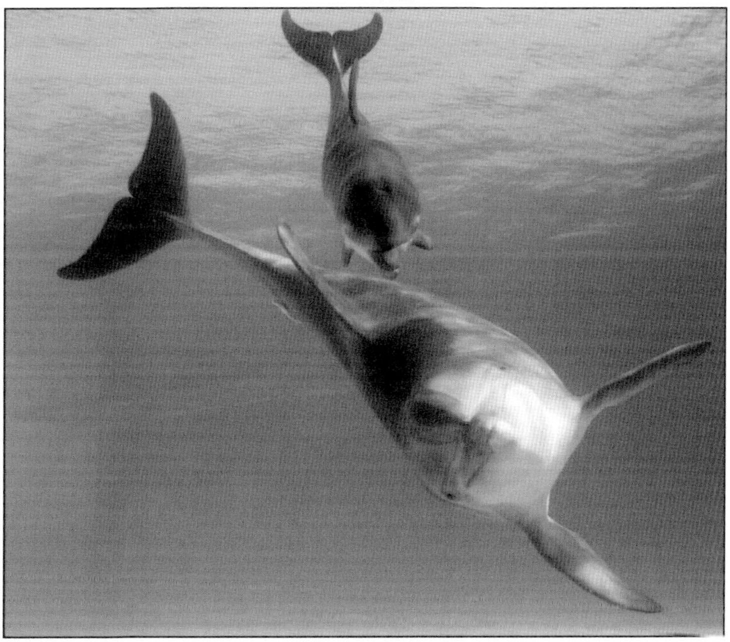

Even media stunts sometimes produced good science. In 1940, Robert L. Ripley, famed for his *Believe It Or Not* radio show, visited Marine Studios and successfully milked a dolphin. The staff actually collected milk from three bottlenose dolphins and sent it to the University of Chicago for analysis. The results, titled "The Composition of Dolphin Milk," were published in 1940.

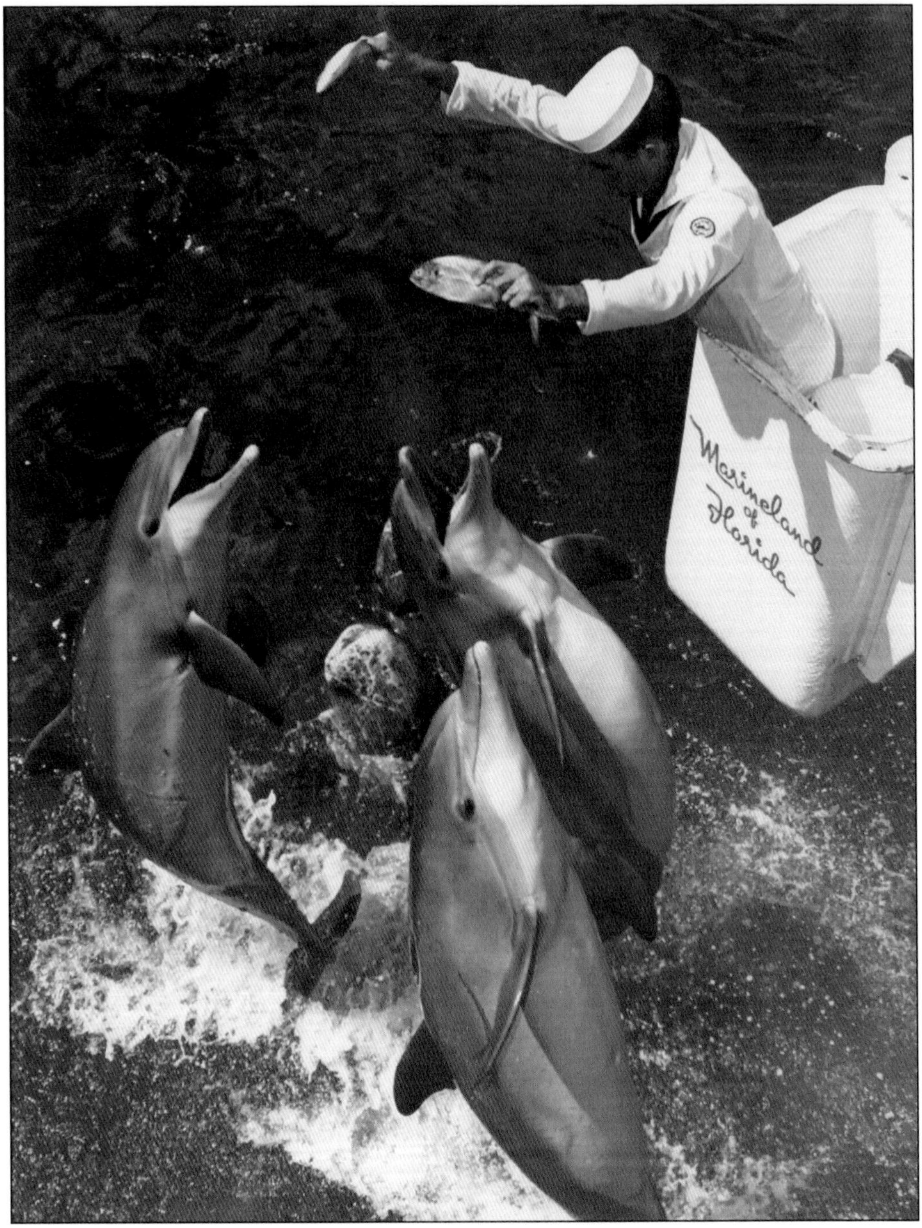

After years of close interaction with dolphins in the circular oceanarium, both the diving staff and jumpmasters began to suspect that the dolphins' level of intelligence and their ability and capacity for learning might be something of interest. Though not animal trainers themselves, the divers had succeeded in creating underwater routines with the dolphins based around mealtime, and the jumpmasters had created a reliable, world-renowned jumping-dolphin show based on the animals' willingness to leap out of the water for their food. Other species of animals, primarily those who lived on land, were commonly trained and seen performing in circuses and at zoos, but the thought of training a creature from the ocean, so entirely different, seemed daunting. A plan began to develop, purely to answer the question, "Can a dolphin be trained?" Marine Studios, a facility steeped in research and exploration, devised a new, interesting, and secretive experiment involving a bottlenose dolphin.

Five

TRAINERS

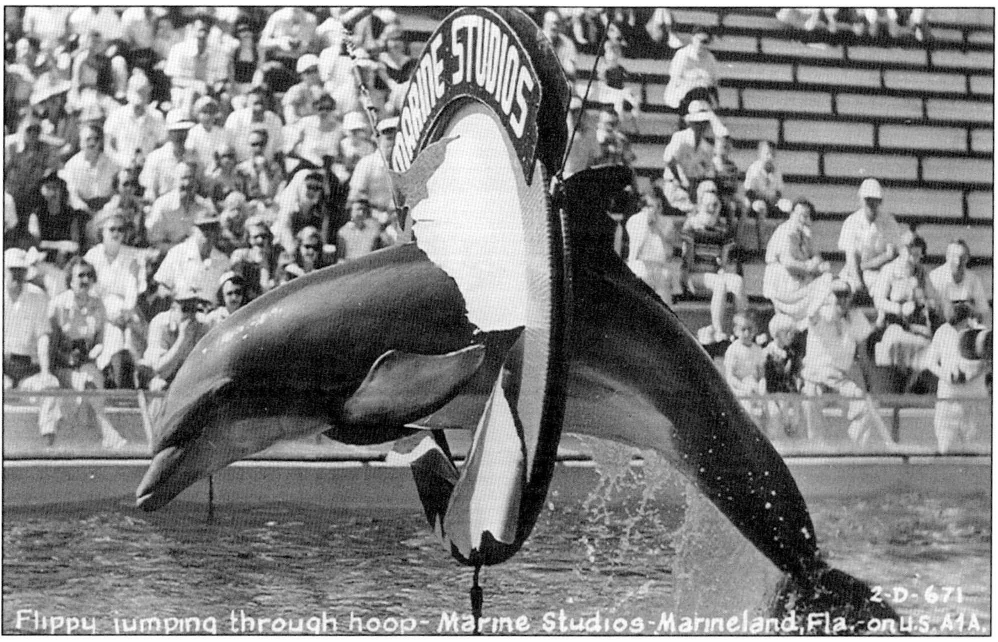

Animal trainers in the 1940s were primarily thought of in the circus ring, subduing ferocious jungle animals with whips and chairs. But witnessing the relative intelligence of the dolphins in their care caused the Marine Studios staff to wonder whether a dolphin could be taught to perform specific tasks on cue. Was a dolphin smart enough to be trained? The world was about to find out.

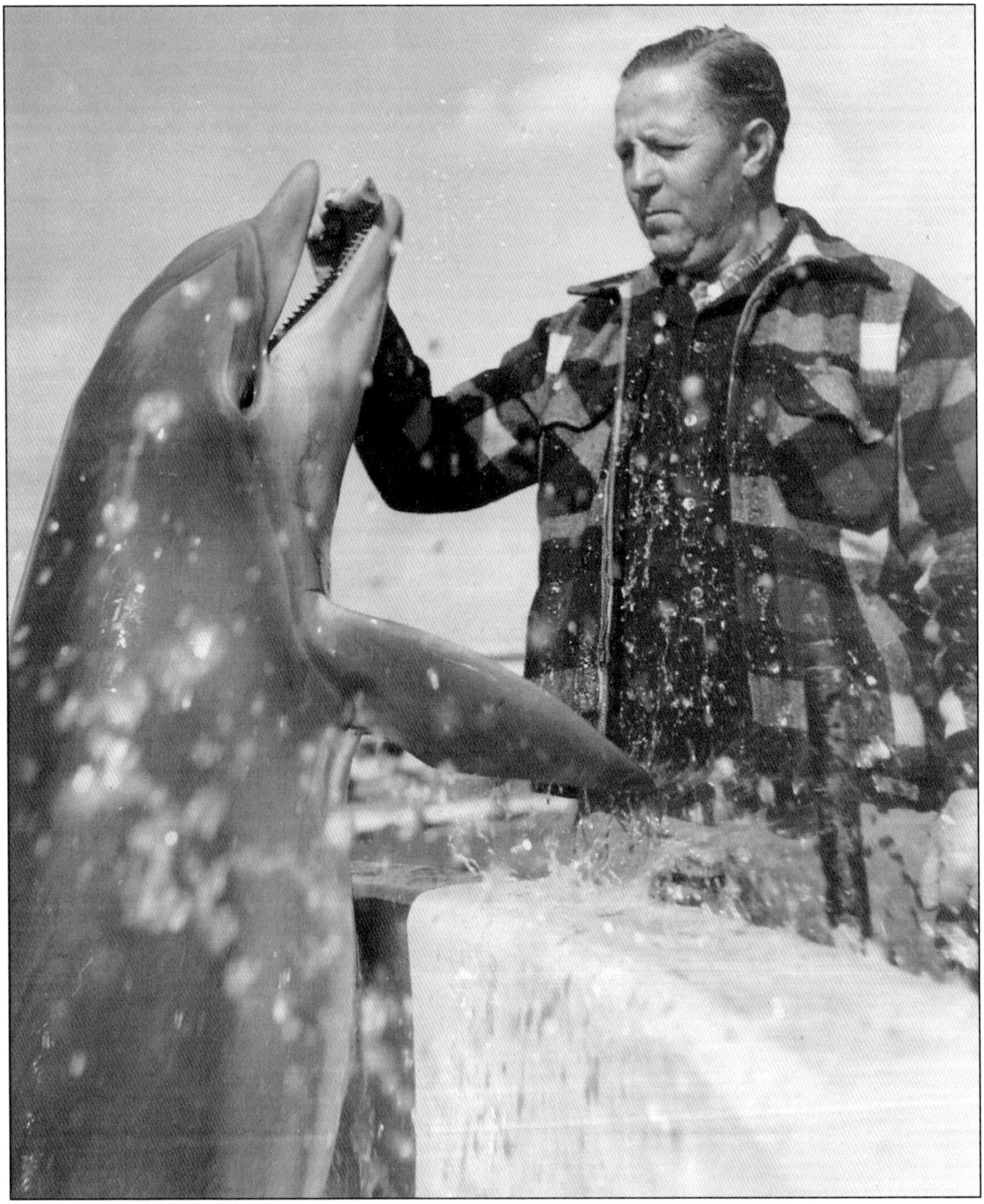

In April 1949, a top-secret dolphin training experiment began, led by Adolf Frohn (shown here), a fourth-generation circus trainer. Frohn had performed in his father's trained sea lion act at the age of 12, worked with world-renowned animal trainer Frank Buck, and was working for Ringling Bros. and Barnum & Bailey Circus at the time. At age 45, when he accepted the position with Marine Studios, Frohn already had 31 years of animal-training experience. Although Frohn was skeptical that a "fish" could be trained to do tricks, he readily accepted the challenge, even though he had never even seen a dolphin until the first day he reported to work. In September 1949, a young male bottlenose dolphin was brought in by the Marine Studios collecting crew. Frohn had requested a young female dolphin since he adamantly believed females of any species learned behavior more quickly than their male counterparts. Disappointed, he reluctantly accepted the challenge of this young male dolphin named Flippy.

Frohn's training style was basic, but sound. His techniques were based on understanding the natural abilities and limitations of the animal while building a trusting relationship that was motivated through encouragement. Frohn believed animal health was paramount and that a hungry animal could not think clearly, so full nutrition was provided daily. Frohn spent his first weeks observing the existing dolphins, teaching himself about them before the experiment began.

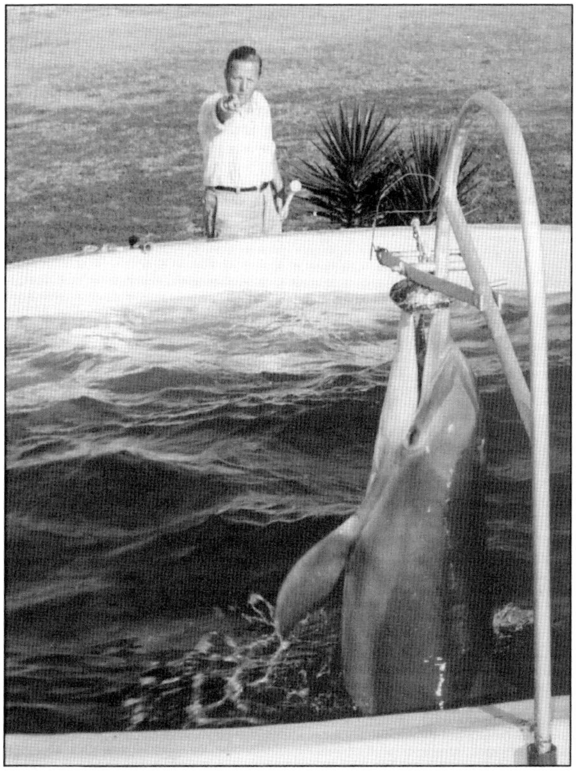

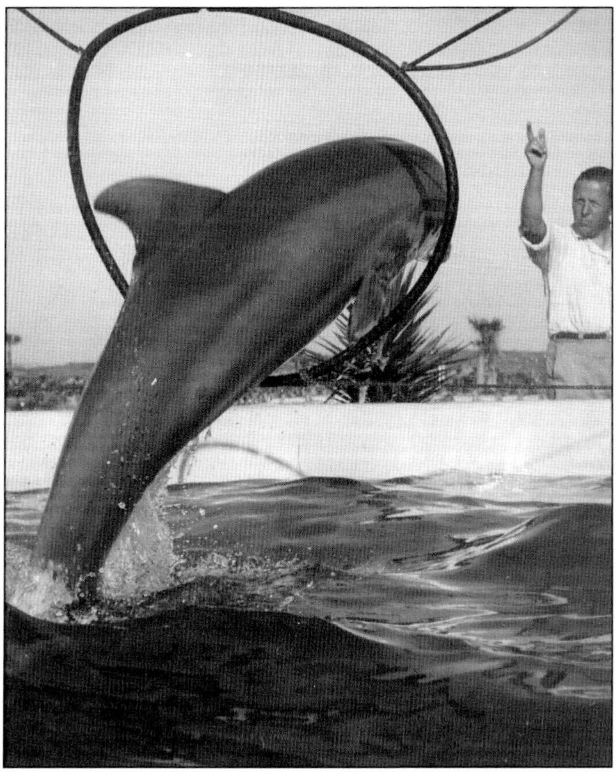

Although a dolphin's ears are often underwater, Frohn was accustomed to training land animals and therefore used his voice almost exclusively to direct, encourage, and reward all at the same time. He paired his voice with a generalized point, used in all cases, to orient an animal in the desired direction. Frohn would also use a pat on the head or a rubdown, saying, "Good, good!" or "Nice boy!"

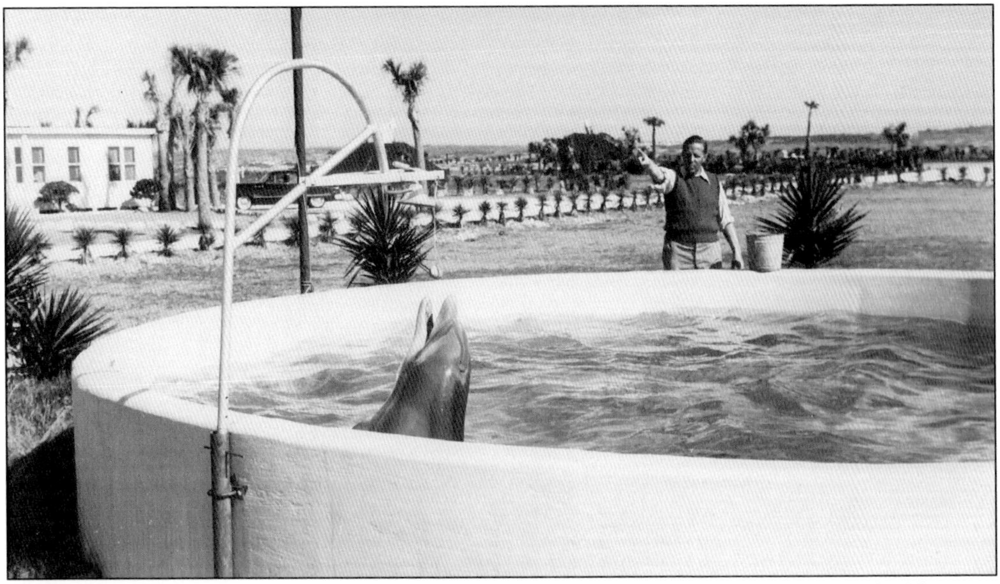

Flippy was housed in a separate facility, appropriately called the experimental pool, hidden in a back area far away from the public's view. The pool had no surrounding structure to limit access, only a sign that read, "Photographs of experiments are restricted." For the next six weeks, Frohn acclimated to his new position and began to build a trusting relationship with Flippy.

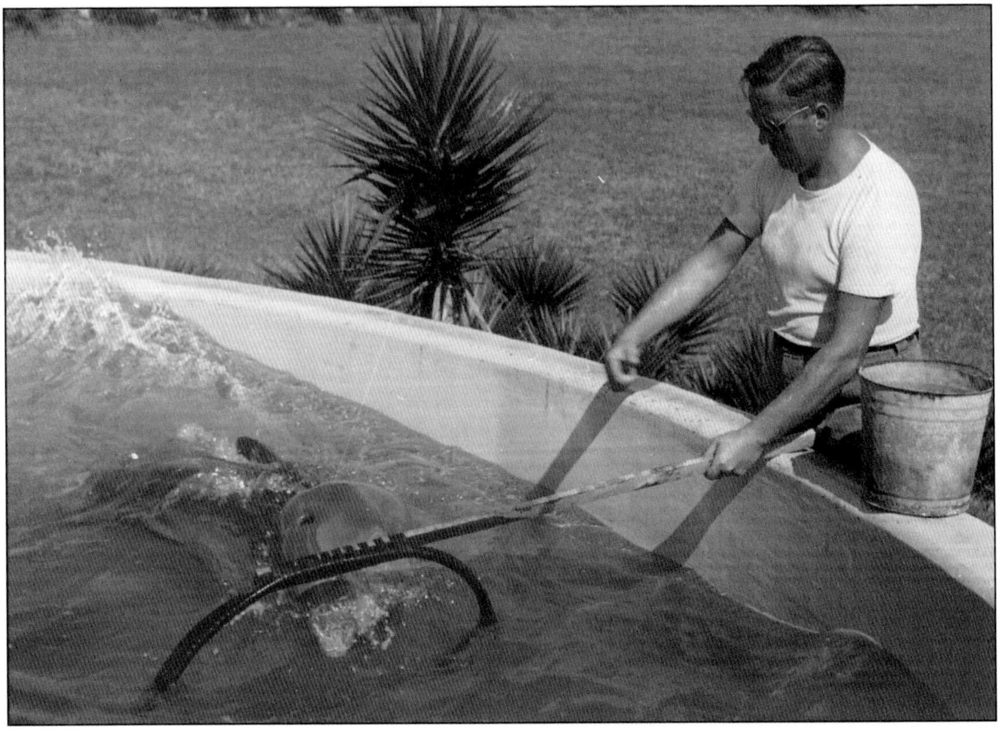

Training began in November 1949, and Flippy's ability to learn exceeded all expectations. He had learned to perform six behaviors in one short year, including jumping through a suspended hoop, jumping to ring a bell, walking back on his tail to catch a ball in his mouth, jumping over a net, spinning in the water, and retrieving a stick.

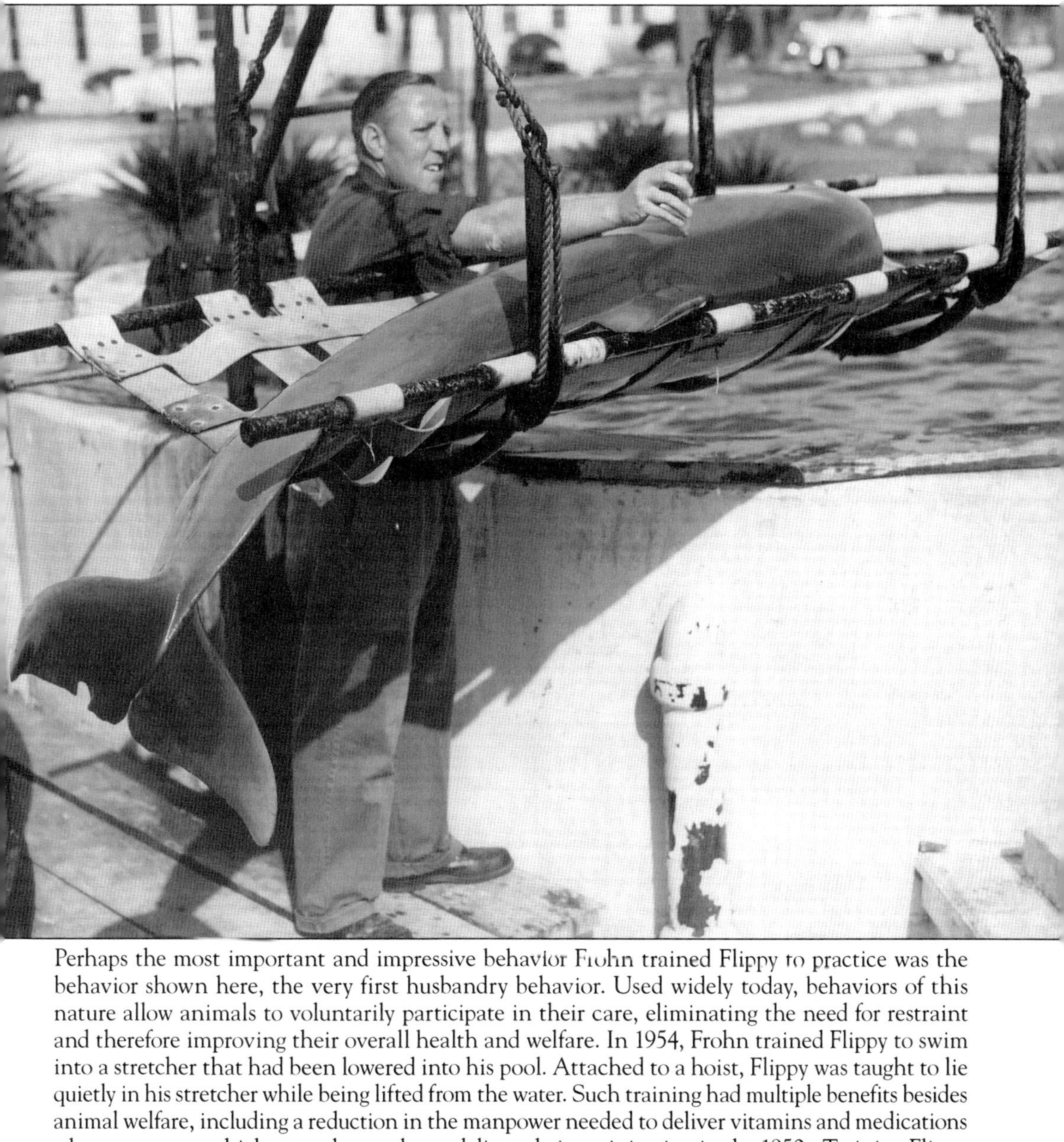

Perhaps the most important and impressive behavior Frohn trained Flippy to practice was the behavior shown here, the very first husbandry behavior. Used widely today, behaviors of this nature allow animals to voluntarily participate in their care, eliminating the need for restraint and therefore improving their overall health and welfare. In 1954, Frohn trained Flippy to swim into a stretcher that had been lowered into his pool. Attached to a hoist, Flippy was taught to lie quietly in his stretcher while being lifted from the water. Such training had multiple benefits besides animal welfare, including a reduction in the manpower needed to deliver vitamins and medications when necessary, which were almost always delivered via an injection in the 1950s. Training Flippy to lie calmly in his stretcher had another unanticipated benefit: It greatly improved the ability to move him to different locations throughout the park. This trained behavior became the model that was desirable for all Marine Studios dolphins and is referred to today as passive restraint.

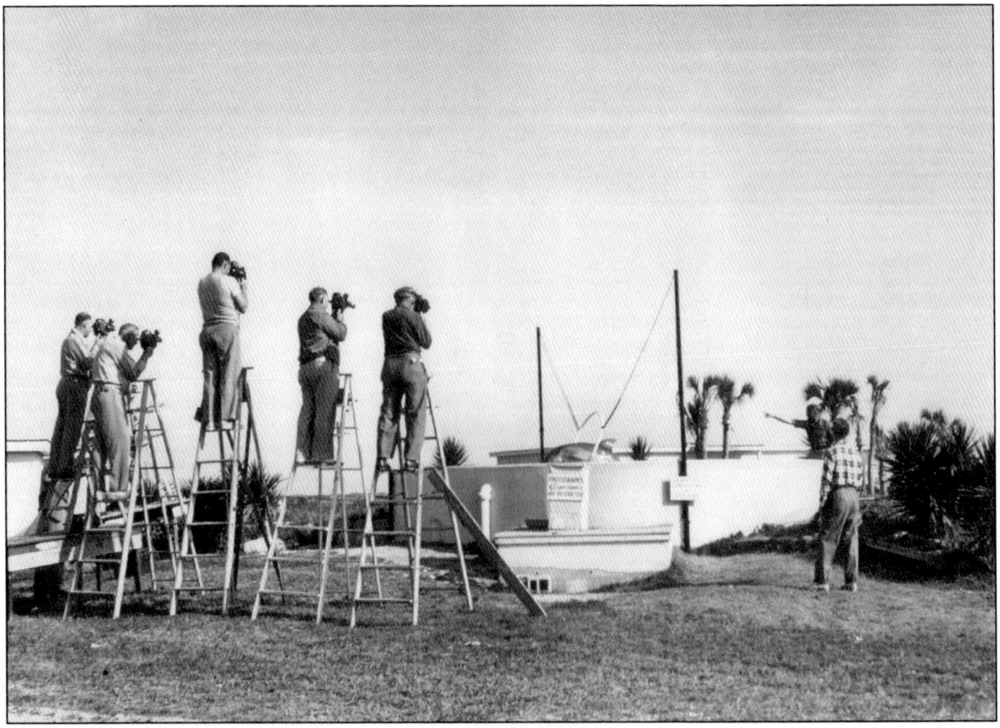

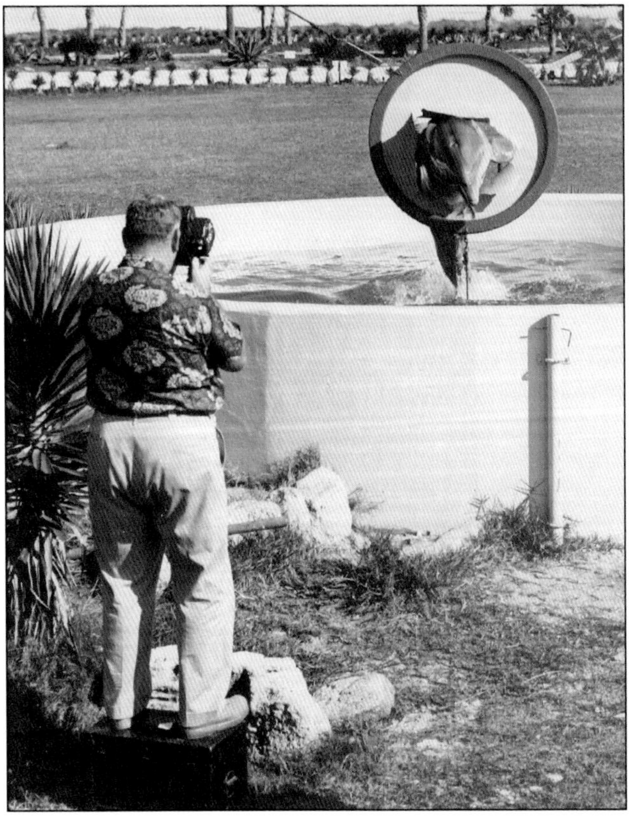

Flippy was introduced to the press on February 28, 1951. His first picture appeared in the *Saturday Evening Post*, making him a national celebrity overnight. In a single month, his picture appeared in over 3,800 newspapers that circulated over 40 million copies. Another 50 million people saw Flippy on newsreels at the movie theater, and television shorts were estimated to reach over 15 million people. Publicity director John Dillin maintained that Flippy was a publicist's dream because Marine Studios didn't have to promote him; instead, people were standing in line, begging to be permitted to publicize him. The enormity of this publicity was not lost on Marine Studios, where private discussions began to formalize the idea of transforming this experiment into a new style of trained-dolphin show for the public.

In March 1951, Flippy was moved to an adjacent lagoon, allowing for a new behavior: pulling a surfboard, as seen here with Pat Dale and Frohn's dog Duke. Released in August 1951, the photograph was accompanied with the admonition that "there are no plans to make the animal part of the Marine Studios exhibit." On exhibit or not, the country had already fallen in love with the world's first trained dolphin.

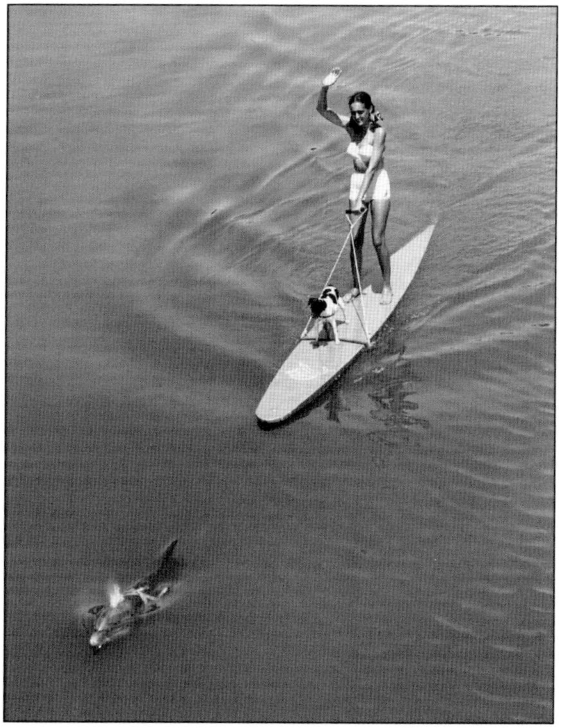

Frohn trained from poolside or a small rowboat because, surprisingly enough, he was unable to swim. Years later, he reflected on the art of dolphin training that he was so instrumental in developing. After devoting his life to the animals in his care, he remained an advocate for what he called the old-fashioned training techniques, including the cornerstones of love and patience.

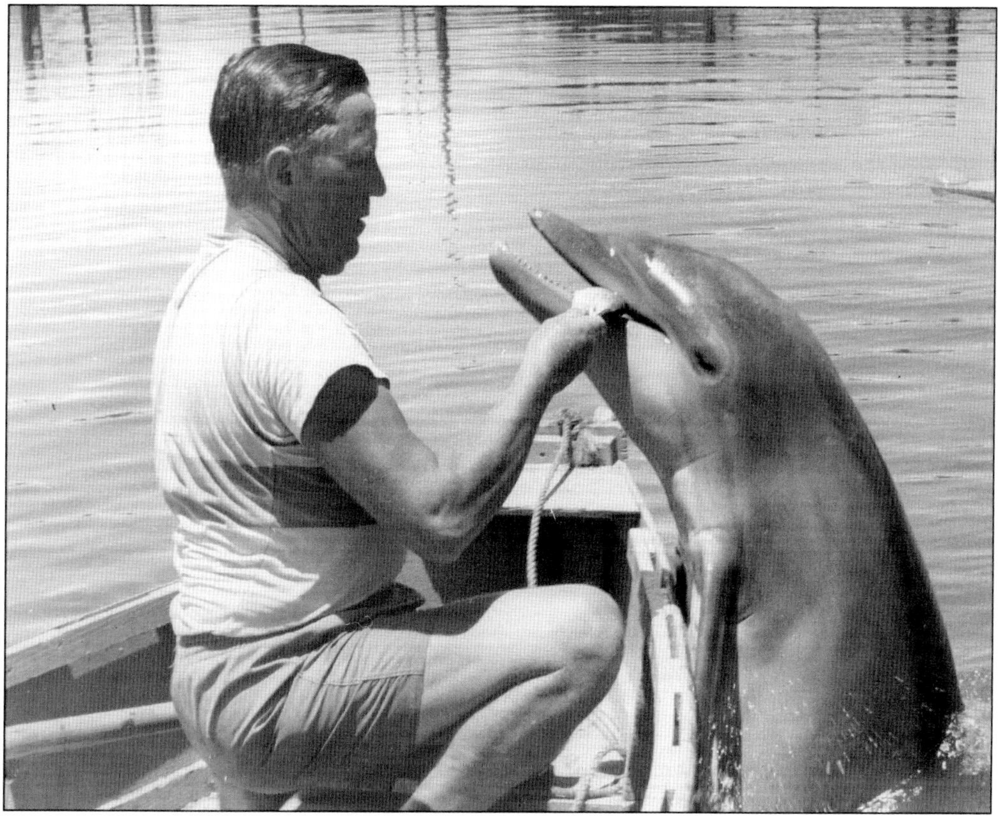

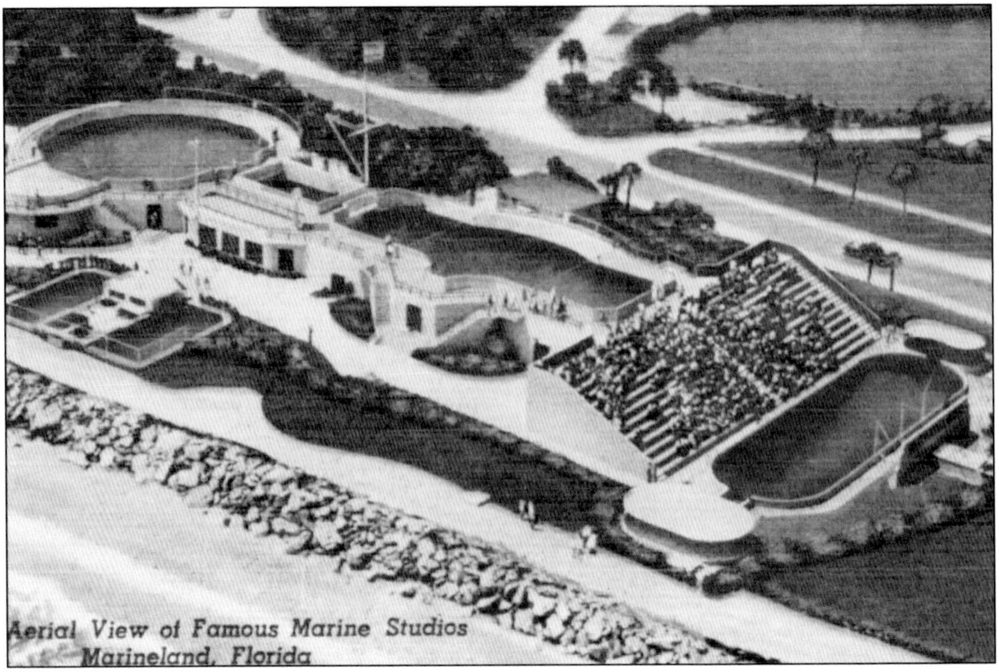

In July 1952, Marine Studios unveiled their exciting and ambitious plans to build a 1,000-seat stadium, where Flippy and understudies Splash and Algae would be featured. With the stadium completed in early 1954, the dolphins were presented in three scheduled training demonstrations a day. They performed only Monday through Friday because Frohn took the weekends off and no one else was trained to work with them. Marine Studios referred to this as an exhibition, not a formal show. The stadium show formally opened to the public in April 1954, but Frohn resigned just before the premiere to begin work at Miami Seaquarium, currently under construction at the time. Frohn's assistants, Hank Czarnota and Andre Cowan, were suddenly promoted to trainers and presented Flippy, Splash, and Algae to an eagerly waiting public in their first shows.

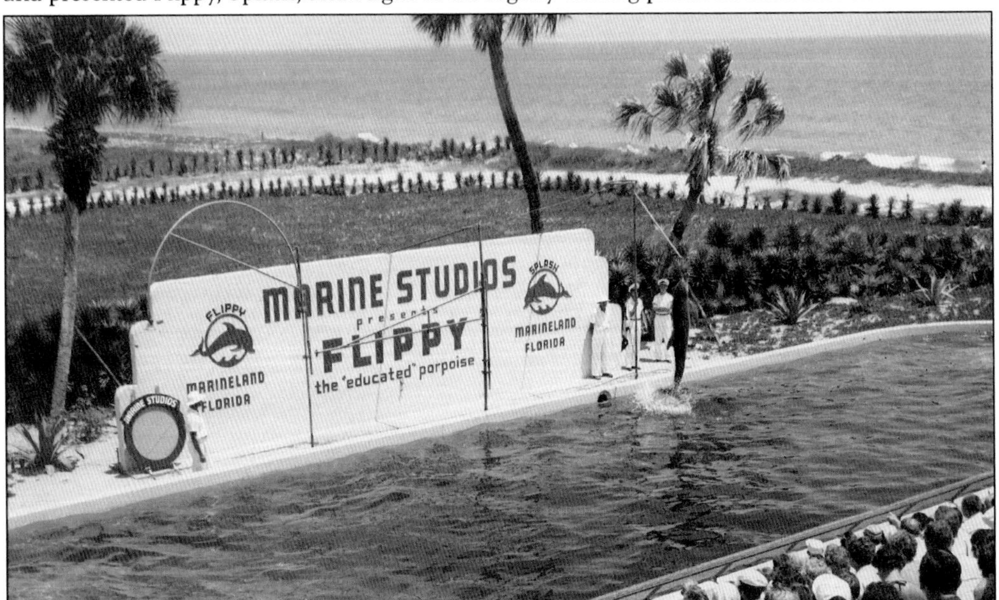

Both Splash and Algae's ability to learn appeared endless, and the trainers were learning a few valuable lessons themselves about dolphin personalities, since there were apparently some real differences between Flippy, Splash, and Algae. Splash excelled at sports, while Algae excelled at high jumping, reaching heights unattainable by any other dolphin in the stadium pool. In August 1955, the existing training program was enhanced when Keller Breland, a contracted behavioral scientist, introduced the idea of using a whistle to signal to the dolphins when they had performed correctly. Marine Studios utilized Breland's expertise for seven years, accompanied by visits from his chief trainer, Kent Burgess, to advance the show. In 1956, Lester Taylor (below left) began his long dolphin-training career at the stadium, being promoted to chief trainer in 1963.

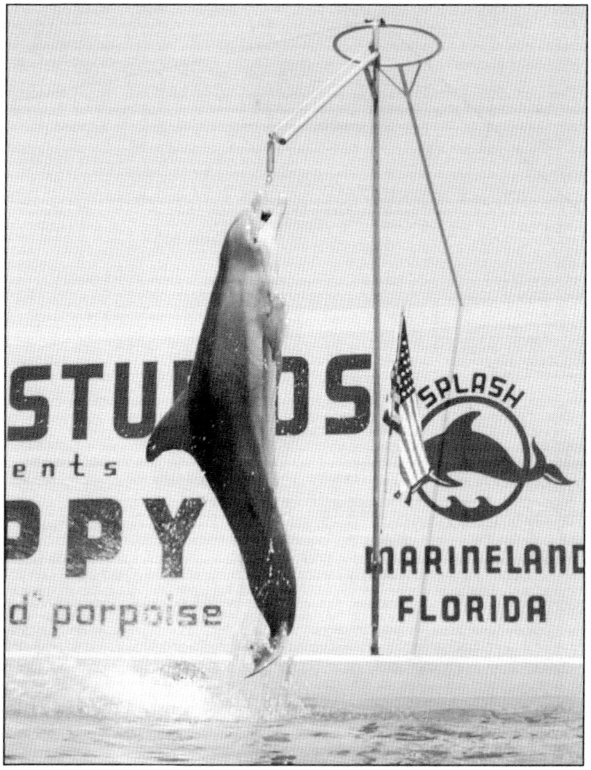

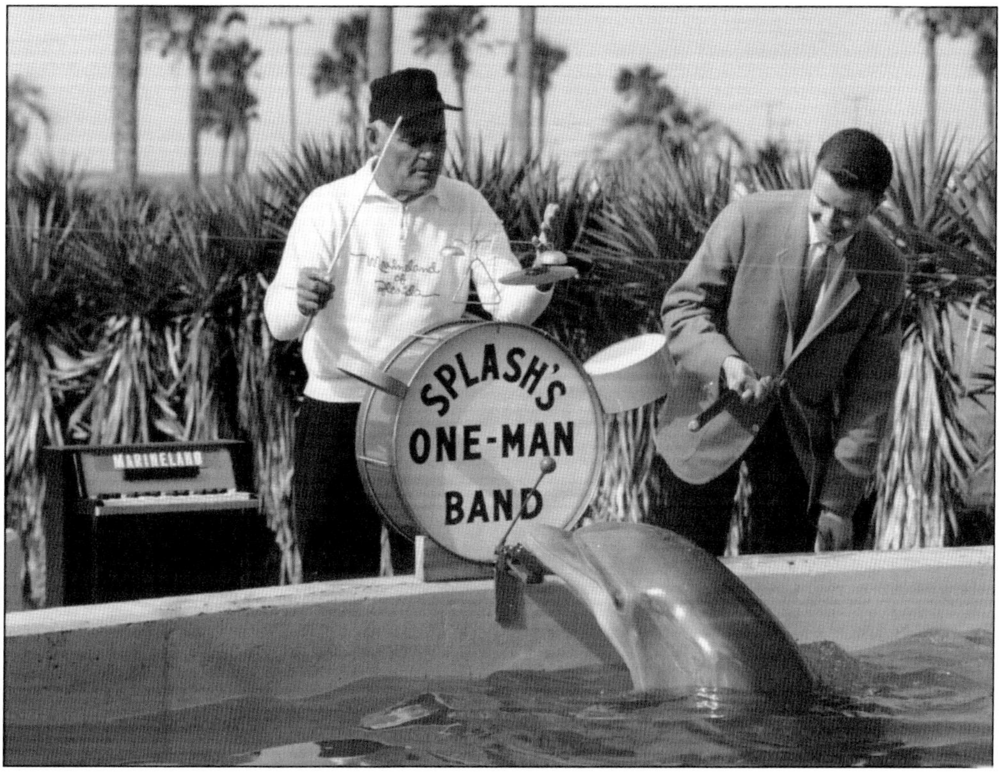

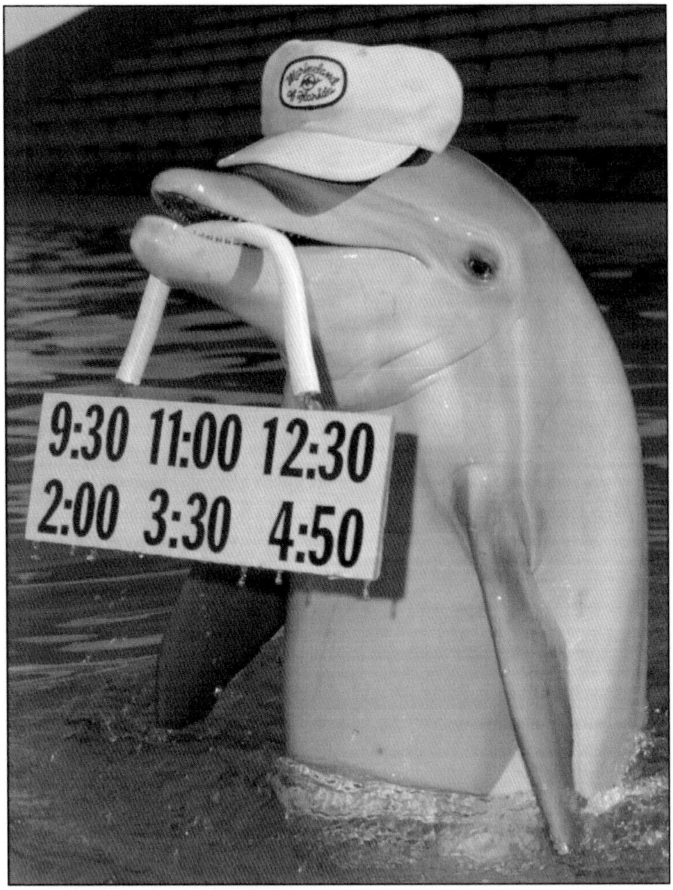

As more and more questions were asked by mystified guests about what they perceived to be the outstanding intelligence of dolphins, Marine Studios put together an answer guide for the inquiring visitors. While the oceanarium agreed that the dolphins had good memories, were capable of recognizing sights and sounds, and possessed some power of reasoning, they hesitated to make any definitive statement about the animals' intelligence. Declaring that scientists had not even managed to agree on the definition of intelligence, and certainly had not yet found a way to measure it in humans, all they could say for sure was the dolphins were trainable.

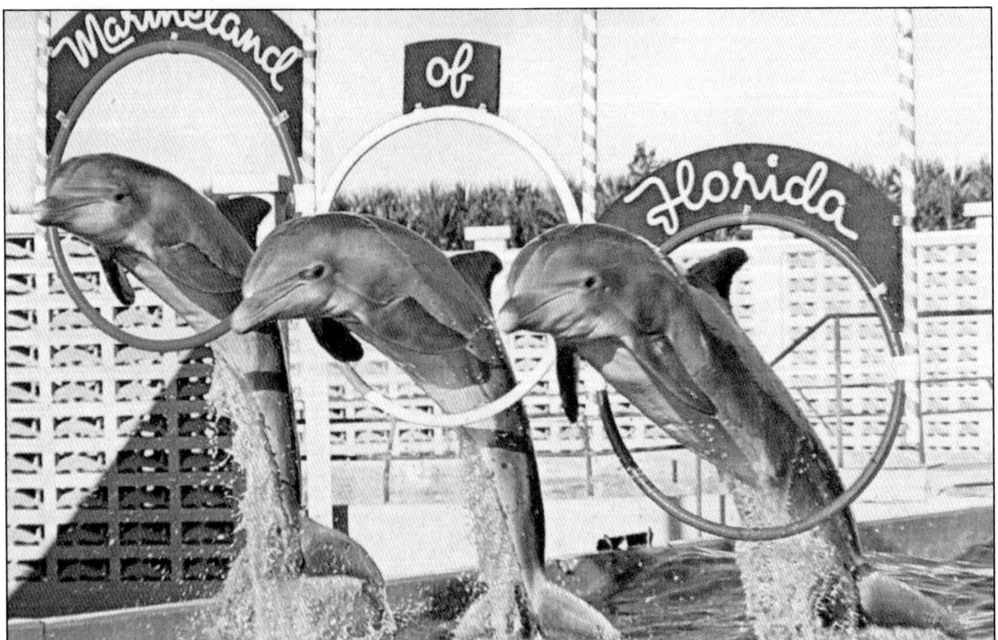

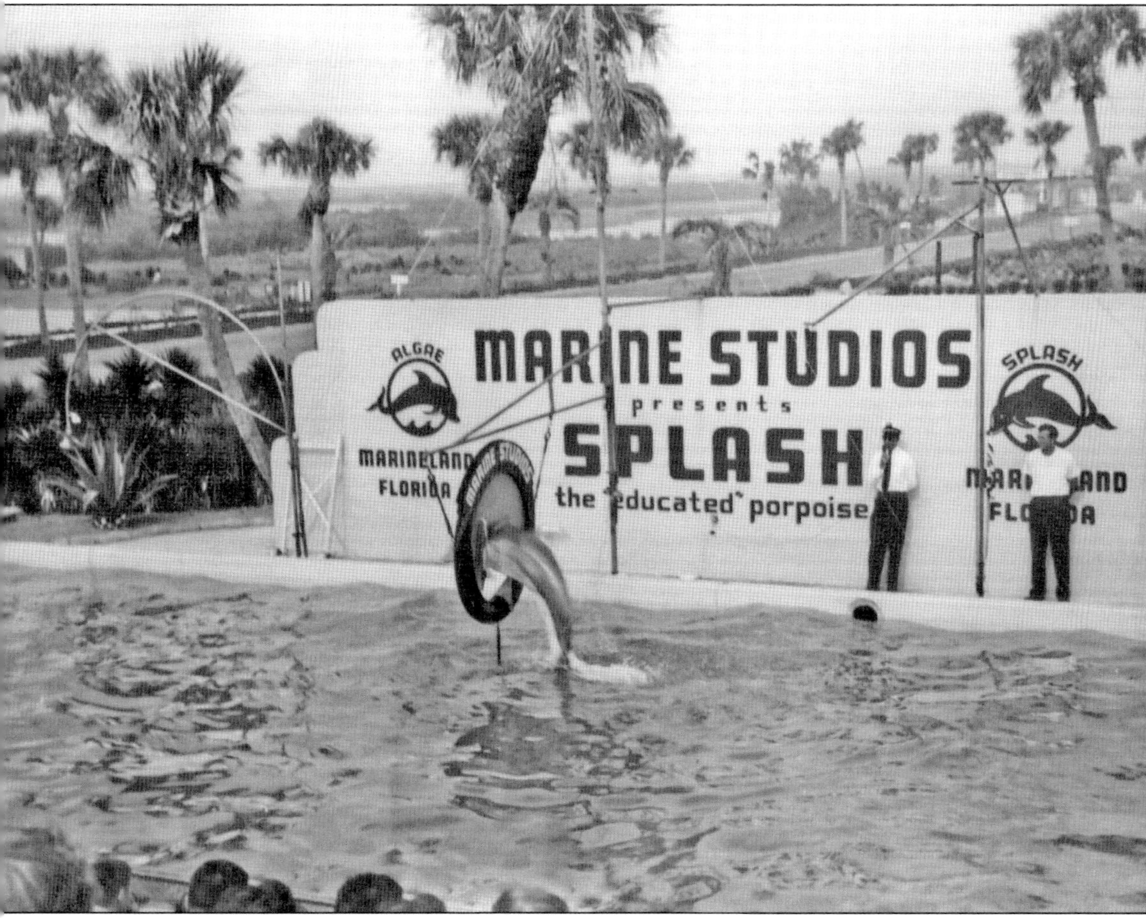

Although advertised as "educated porpoises," none of the performing dolphins were actually porpoises at all. In fact, Marine Studios never displayed porpoises in its animal collection. Despite that fact, the term was used to describe the dolphins for years. The "world's first and only educated porpoises" remained one of the proudest claims made. In fact, publicity director John Dillin used the key phrase to title an article, "Flippy the Educated Porpoise," published in *Natural History* magazine in 1952 and subsequently won an award from the American Association for the Advancement of Science. The article featured several comical observations about Frohn's views on dolphin intelligence. Marveling at everyone's conclusion that Flippy was genuinely intelligent, he wondered how they reached that decision, especially since he himself had little opportunity to compare Flippy's learning abilities to other dolphins; therefore, Frohn concluded that Flippy might actually be quite dumb. Responding to the idea that Flippy would soon be talking if he got any smarter, Frohn wisely commented that Flippy did talk, but we were just not smart enough to understand him.

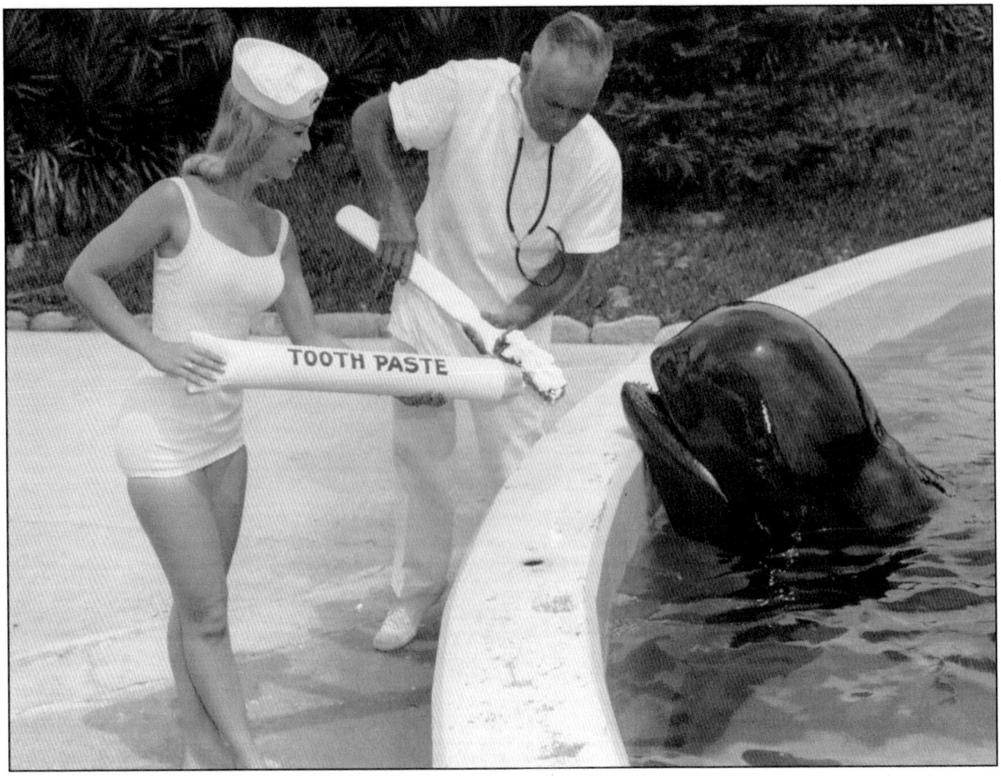

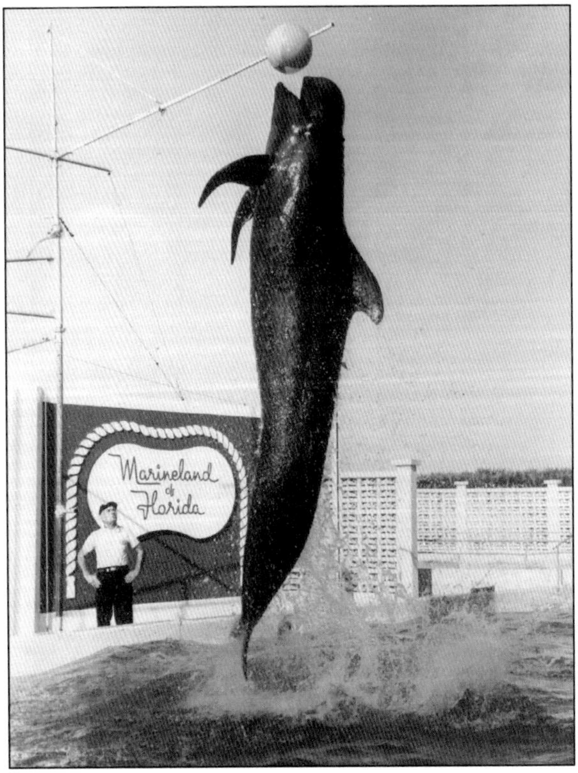

In 1962, sister park Marineland of the Pacific, in Rancho Palos Verdes, California, brought a female pilot whale named Kay to the Florida park. She lived in the show stadium, now boasting "Educated Porpoises and Whales." The show featured a skit in which Kay was examined by the doctor, including a listen through the stethoscope and a thorough teeth cleaning with a whale-sized toothbrush.

In 1969, a male pilot whale named Moby joined Kay in the show pool. It was Moby, also from Marineland of the Pacific, who stole the show with his ability to jump almost 20 feet in the air. At 1,600 pounds, his landings made a huge splash, sending enormous amounts of seawater spilling over the pool walls and into the stadium of cheering spectators.

Six

PERFORMERS

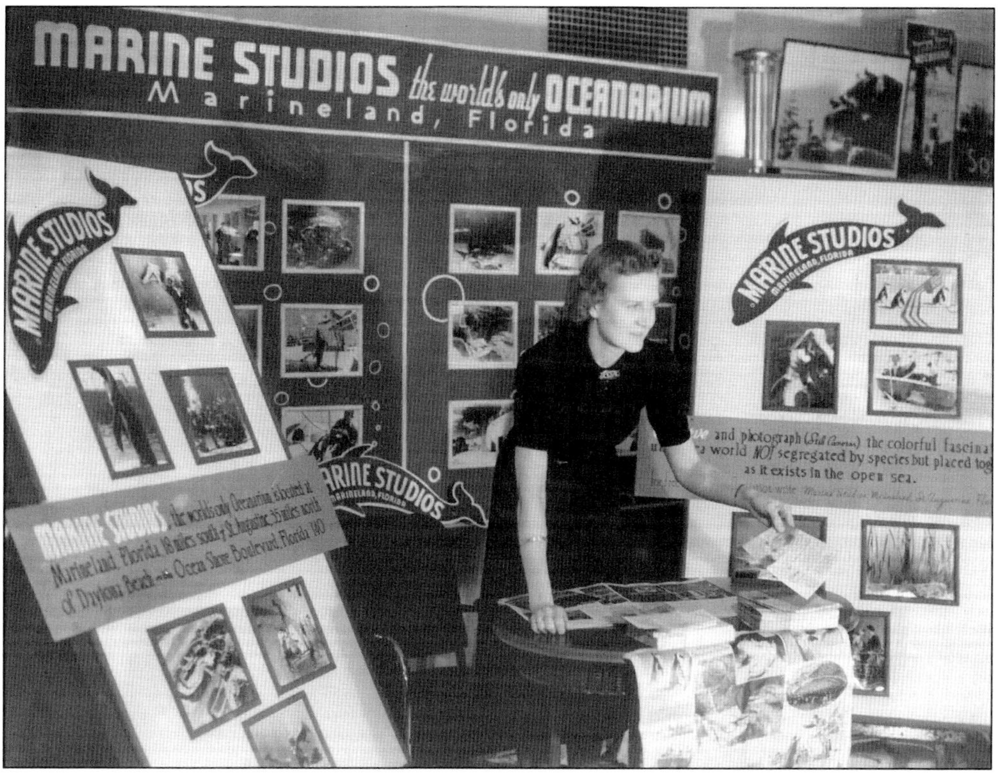

Frequently visited by celebrities and authors, Marine Studios found itself repeatedly in the center of the promotional limelight. Motion picture newsreels, television shows, and feature-length films took advantage of the cinematic opportunities that only Marine Studios could offer, while visiting celebrities almost never left without a photograph, usually featuring themselves as jumpmaster. Although designed as a film studio, the oceanarium was frequently its own biggest star.

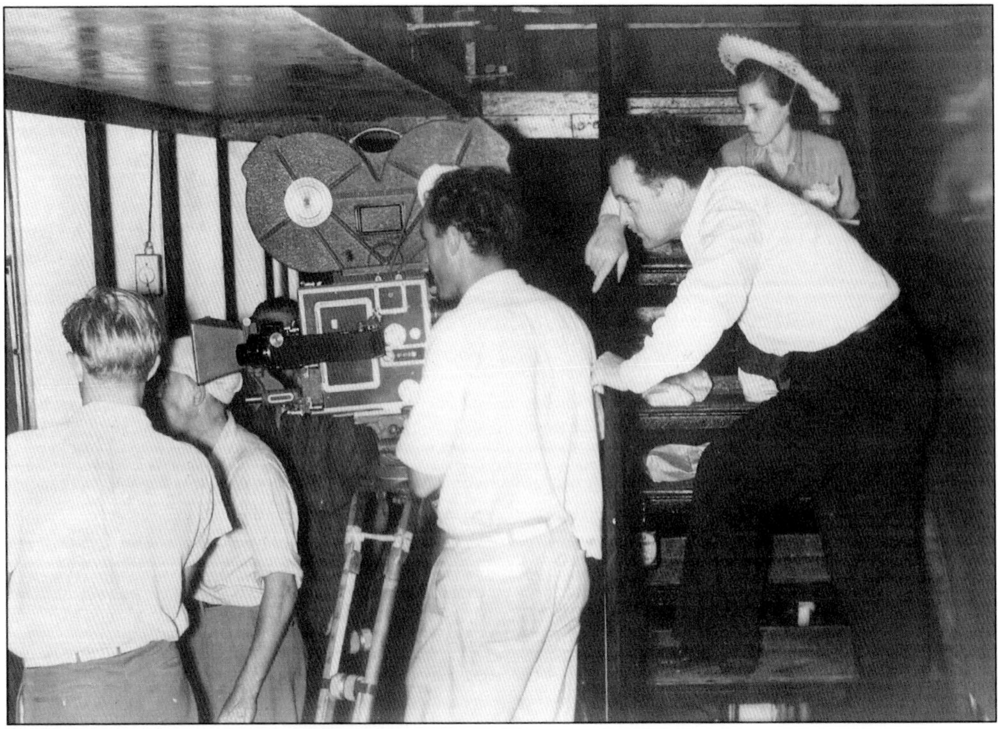

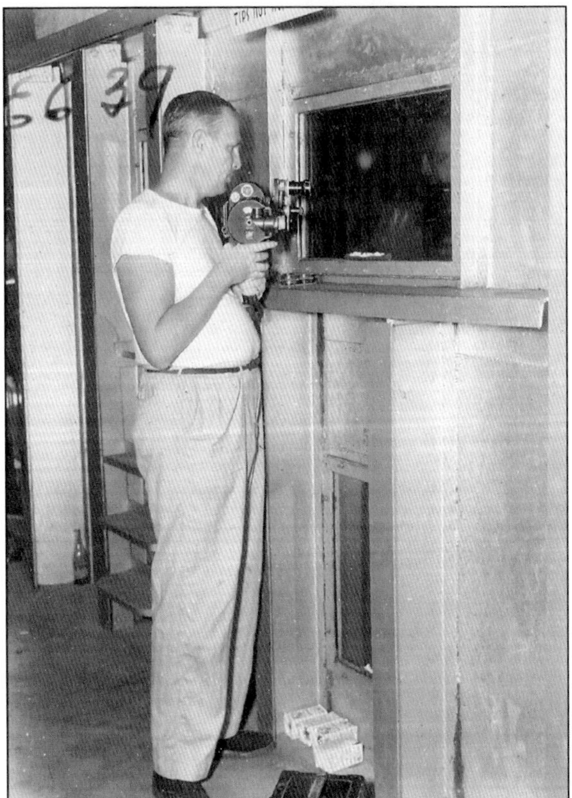

The forethought that went into the design of the oceanarium produced not only a strong structure withstanding the intense inner pressure of close to a million gallons of seawater, but also a structure whose every curve was designed to increase filming opportunities. Over the years, production crews, movie stars, and professional models benefitted from the ideal design and perfect angles offered by Marine Studios. While professional movie cameras were always welcomed, it was not until 1946 that visitors were permitted to bring their 8-mm and 16-mm home movie cameras into the park. The early ban on movie cameras did not apply to still photography cameras, which had never been banned.

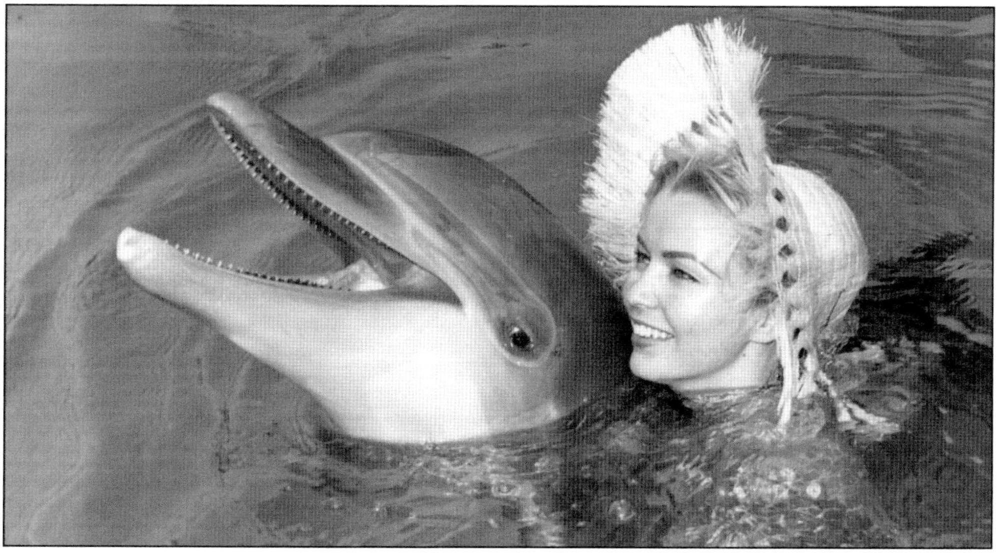

Marine Studios has been proud of its heritage and location, dedicating considerable time and resources to worthwhile community projects throughout its history; however, nothing seemed to outshine the unbeatable combination created by introducing the beautiful creatures of the sea to Florida's beautiful bathing beauties. It was Florida's tourism industry that understood images of dolphins would entice more visitors to the Sunshine State from across the nation. Florida's citrus queen and Pancho the dolphin seem to make an unbeatable pair. In fact, CBS's coverage of Florida's Citrus Exposition in 1957 featured Marine Studios' dolphins as an added attraction. The exposition was a weeklong event promoting the Florida citrus industry in a state fair style, featuring exhibits, auctions, a midway, and prize-winning fruit, while enticing Americans around the world with the allure of sunny Florida.

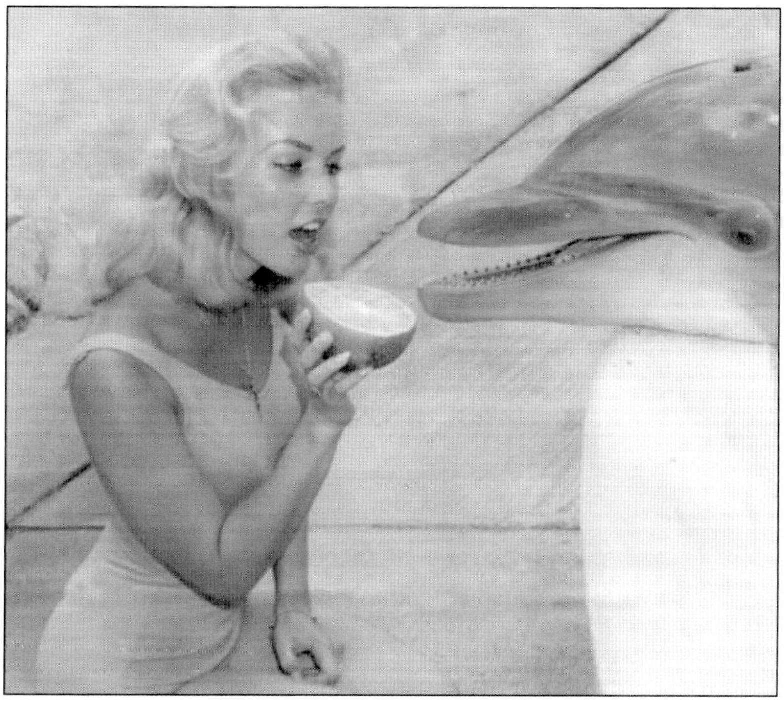

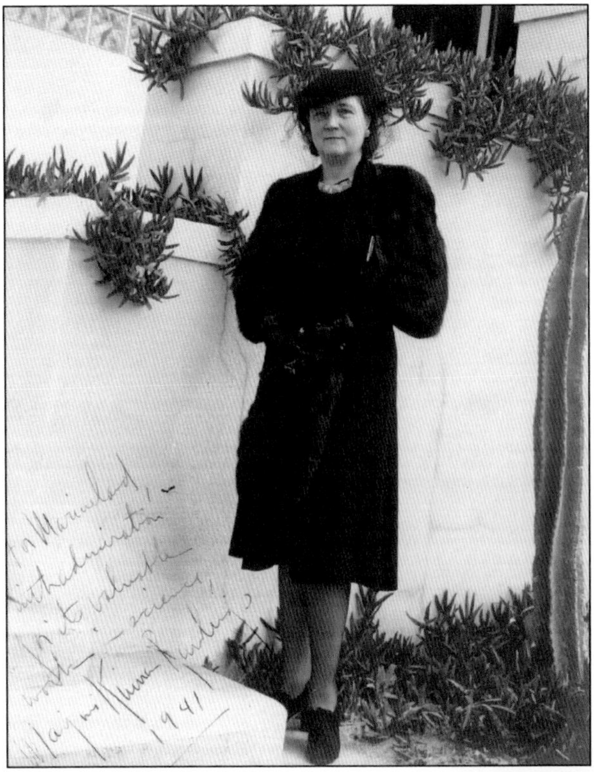

Among the celebrities known to frequent Marine Studios were many great authors. Often recognized by their names and not their faces, they vacationed along the Marineland coast completely unnoticed. As one story goes, in 1940, Marjorie Kinnan Rawlings (left), author of *The Yearling*, was sitting in the Dolphin Restaurant with her soon-to-be husband Norton Baskin, owner and operator of the restaurant, when she thought she recognized someone sitting across the room. She decided to send over a note saying, "If you're Ernest Hemingway, won't you join us for a drink?" He reportedly wrote back, "If you're Marjorie Rawlings, I'd be glad to." Hemingway (below) grew to be a friend of the couple, spending time with them at Marineland and at their nearby beachfront cottage.

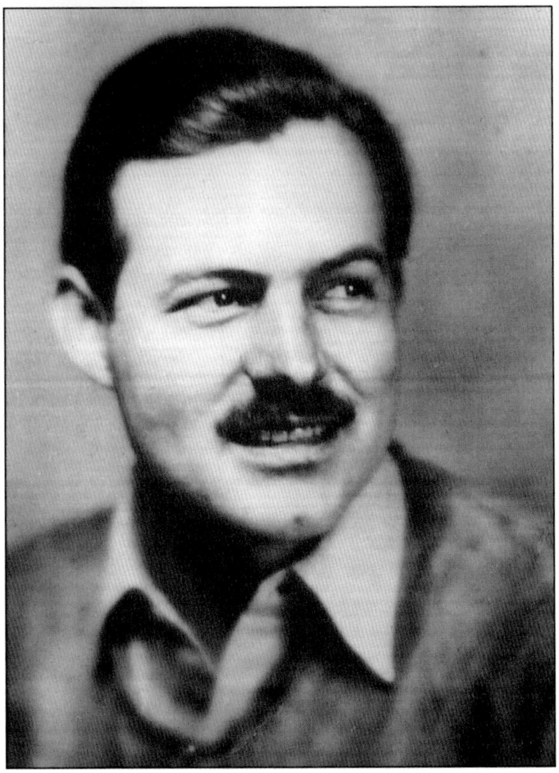

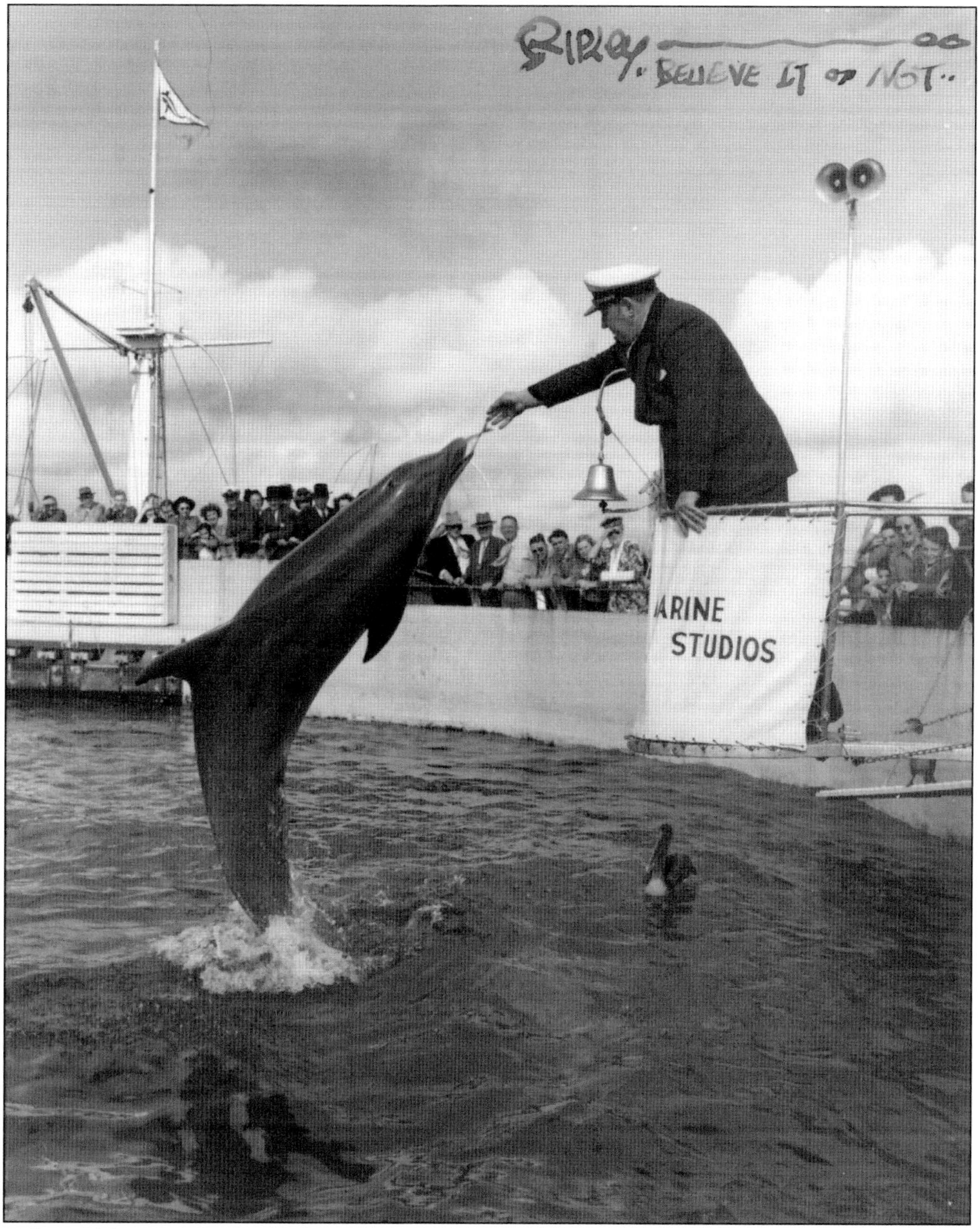

Other celebrities to visit Marine Studios included Robert L. Ripley, famed for his *Believe It Or Not* series, who conducted the first coast-to-coast underwater radio broadcast, featuring an interview with diver Bill Zeiler from the bottom of the Marine Studios pool. The special was broadcast to all CBS affiliates across America on February 23, 1940. There were an estimated 30 million listeners tuned in. During his four-day visit, Ripley himself donned the diving helmet to hand-feed the dolphins, sharks, turtles, and fish while broadcasting live from the bottom of the pool. Underwater broadcasts of this type soon became frequent and common occurrences in the rectangular oceanarium, with divers talking through an RCA lapel microphone while wearing a set of headphones, each secured inside their dive helmets. As seen here, Ripley took time to make friends with the dolphins from above the water as well.

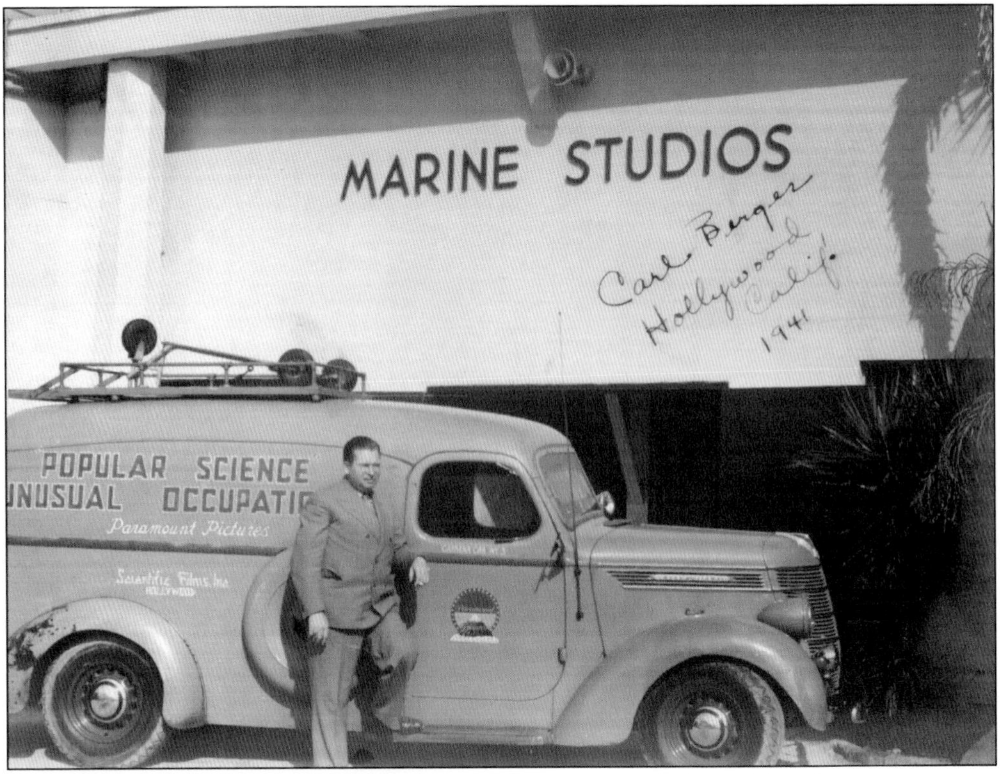

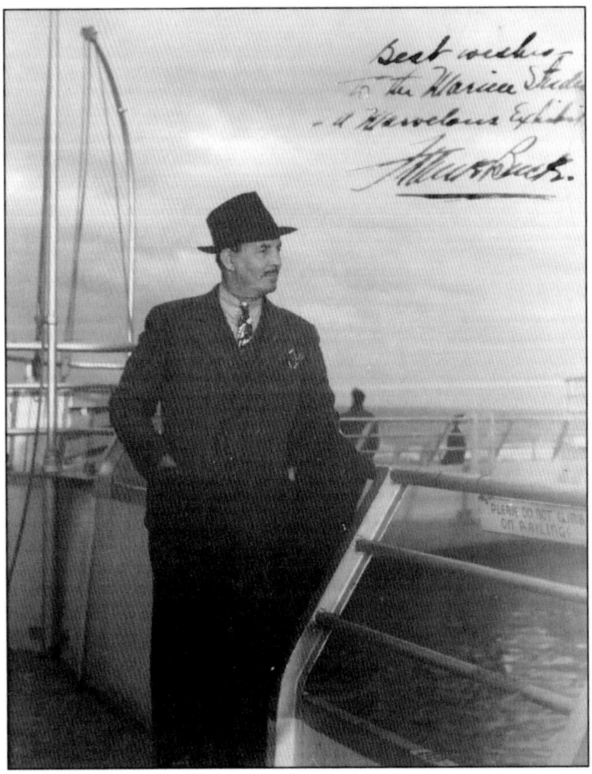

Carl Berger (above) was a cinematographer in the 1940s, known primarily for his involvement in the production of *Bring 'Em Back Alive*, the film portrayal of the real life of animal collector Frank Buck. Shown here with a note congratulating Marine Studios on what he called a "marvelous exhibit," Frank Buck was responsible for collecting many of the animals displayed in zoos and circuses in the early 1900s. In between collecting expeditions, Buck was involved in the display of animals himself and had his own exhibit at the 1939 World's Fair in New York, which he called Frank Buck's Jungleland. Familiar with jungle animals exclusively, Buck was not involved in the collection of specimens for Marine Studios, but was as awed by the marine animals as any other guest.

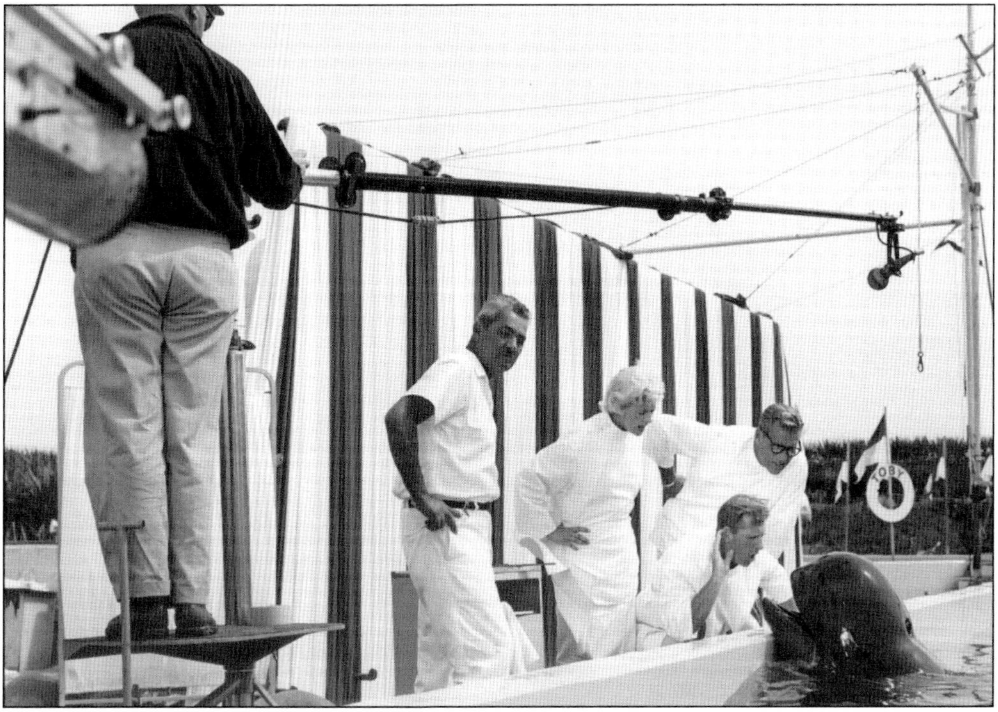

Other celebrity visitors to Marine Studios included Lloyd Bridges and Rosemary Clooney, who hosted *Marineland Circus* in 1961. Seen above, Bridges serves as Toby the pilot whale's physician while Clooney and others look on. Bridges, well known for his role as Navy frogman Mike Nelson in the television series *Sea Hunt*, filmed portions of that show at Marine Studios as well.

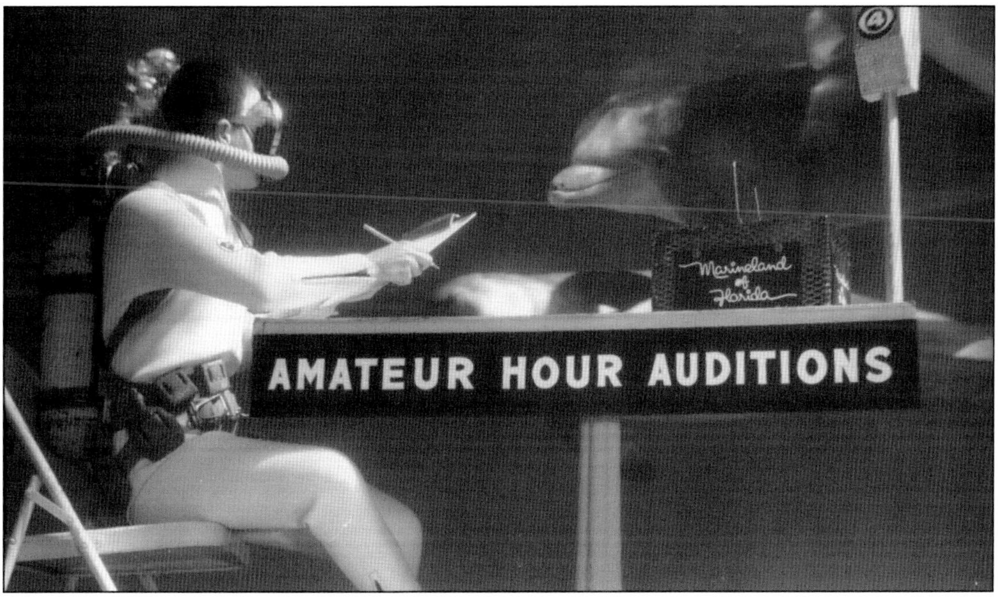

Ted Mack's *Original Amateur Hour* usually required contestants to show up at New York's Radio City Music Hall to audition for the show, but an exception was made for the residents of the circular oceanarium, who were auditioned on their home territory. Here, several bottlenose dolphins wait their turn to be a part of this television show that aired on all three major networks in the early 1950s.

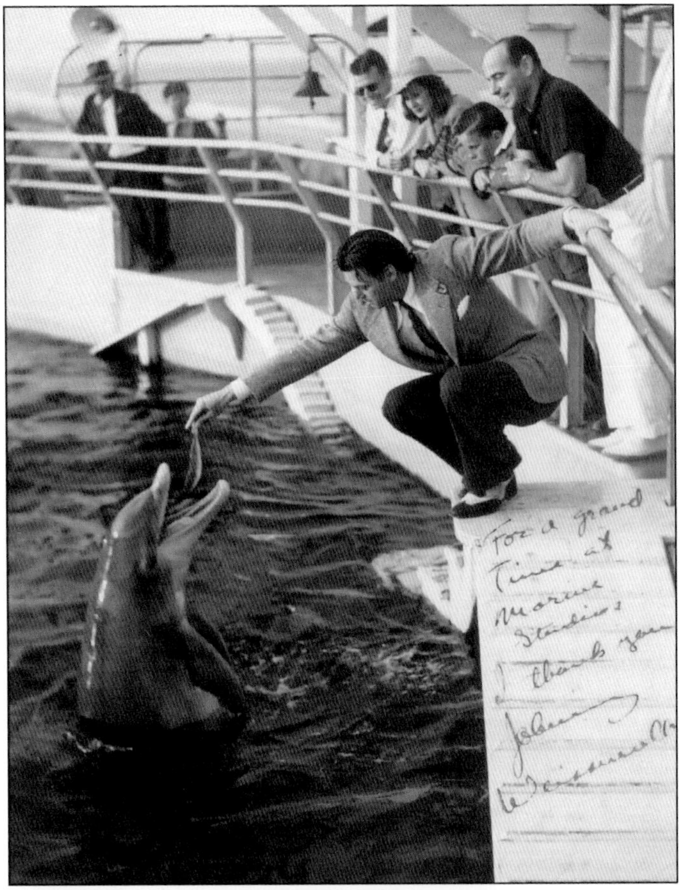

Also filmed at Marine Studios were portions of the Tarzan films of the 1940s. Johnny Weissmuller, Olympic swimmer turned actor, starred as Tarzan in six of the films. His swimmer's physique and unique ability to ululate, producing the notorious Tarzan call, helped make him famous. Weissmuller traveled the world to film *Tarzan*, *Tarzan the Ape Man*, *Tarzan Escapes*, *Tarzan and His Mate*, *Tarzan Finds a Son*, *Tarzan's Secret Treasure*, and *Tarzan's New York Adventure*, but the set at Marine Studios and in the surrounding tropical habitat was said to be his favorite. In this autographed photograph, Weissmuller thanks Marine Studios "for a grand time" while he feeds a dolphin in the circular oceanarium with founder Ilia Tolstoy looking on.

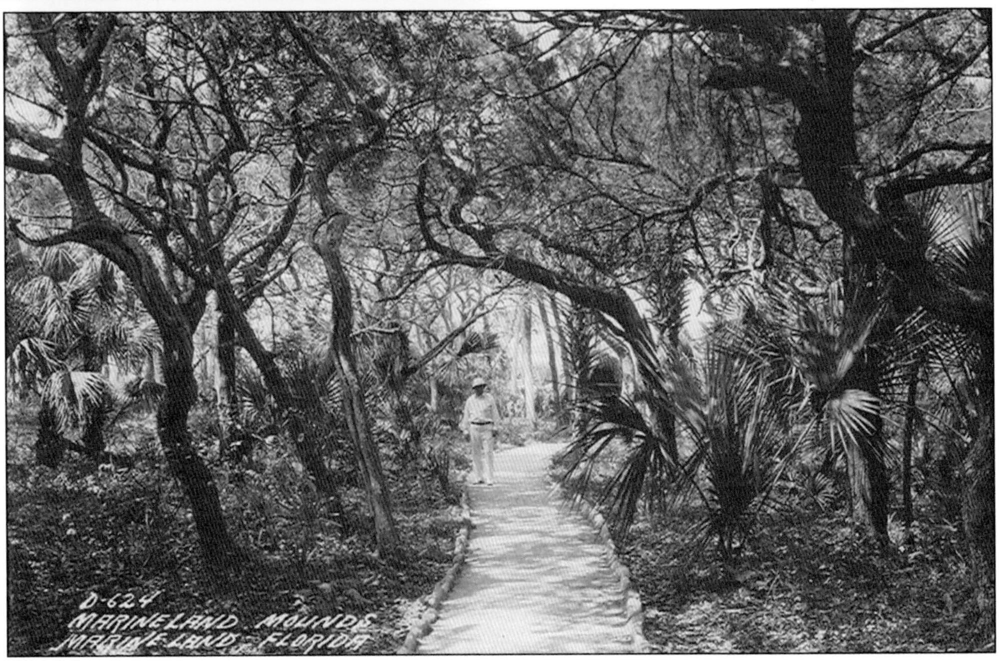

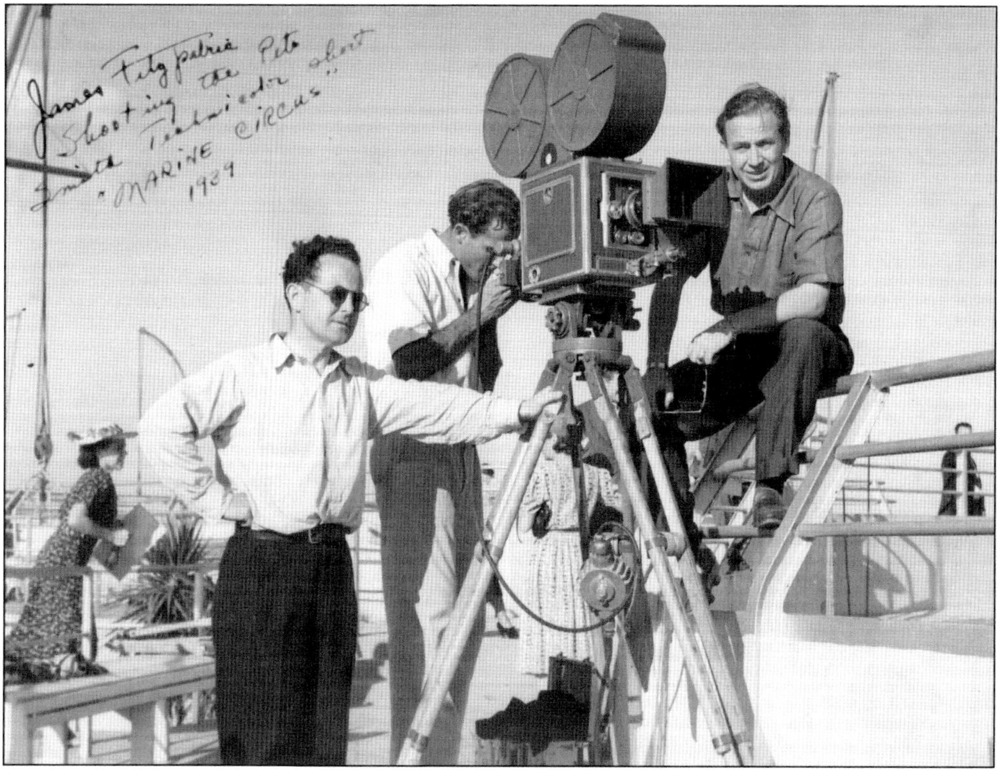

Newsreels and short films were a wonderful way for Marine Studios to keep itself in the public eye. As early as December 1938, the Matanzas Theatre in St. Augustine featured a short Technicolor film by Pete Smith, focusing on the creatures of Marine Studios, titled *Marine Circus*. It was subsequently released to 20,000 theaters around the country and shown prior to feature films.

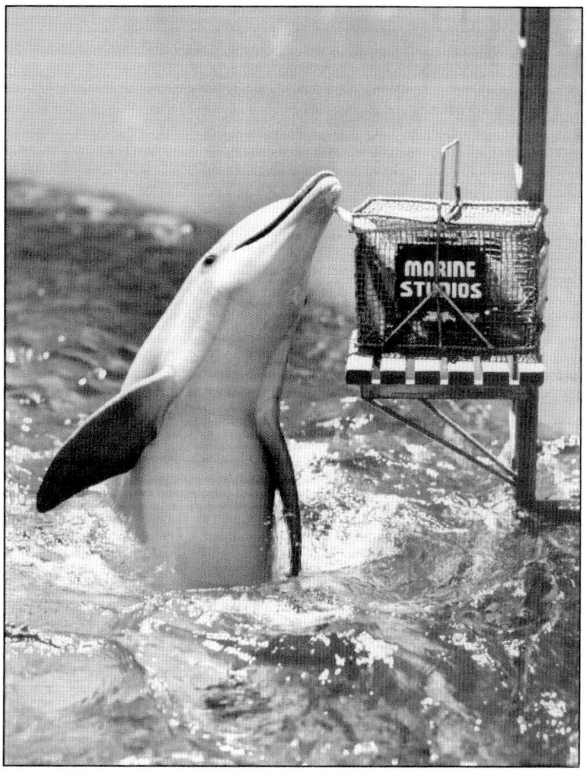

Newsreel groups often visited Marine Studios, but by far the most popular and successful newsreel of 1941 was *Fish Census*, combining underwater performances of beautiful mermaids from Weeki Wachi Springs in the oceanarium with Marine Studios' fascinating creatures from the deep. Due to its enormous success, the fish census idea became the subject of annual filming following World War II.

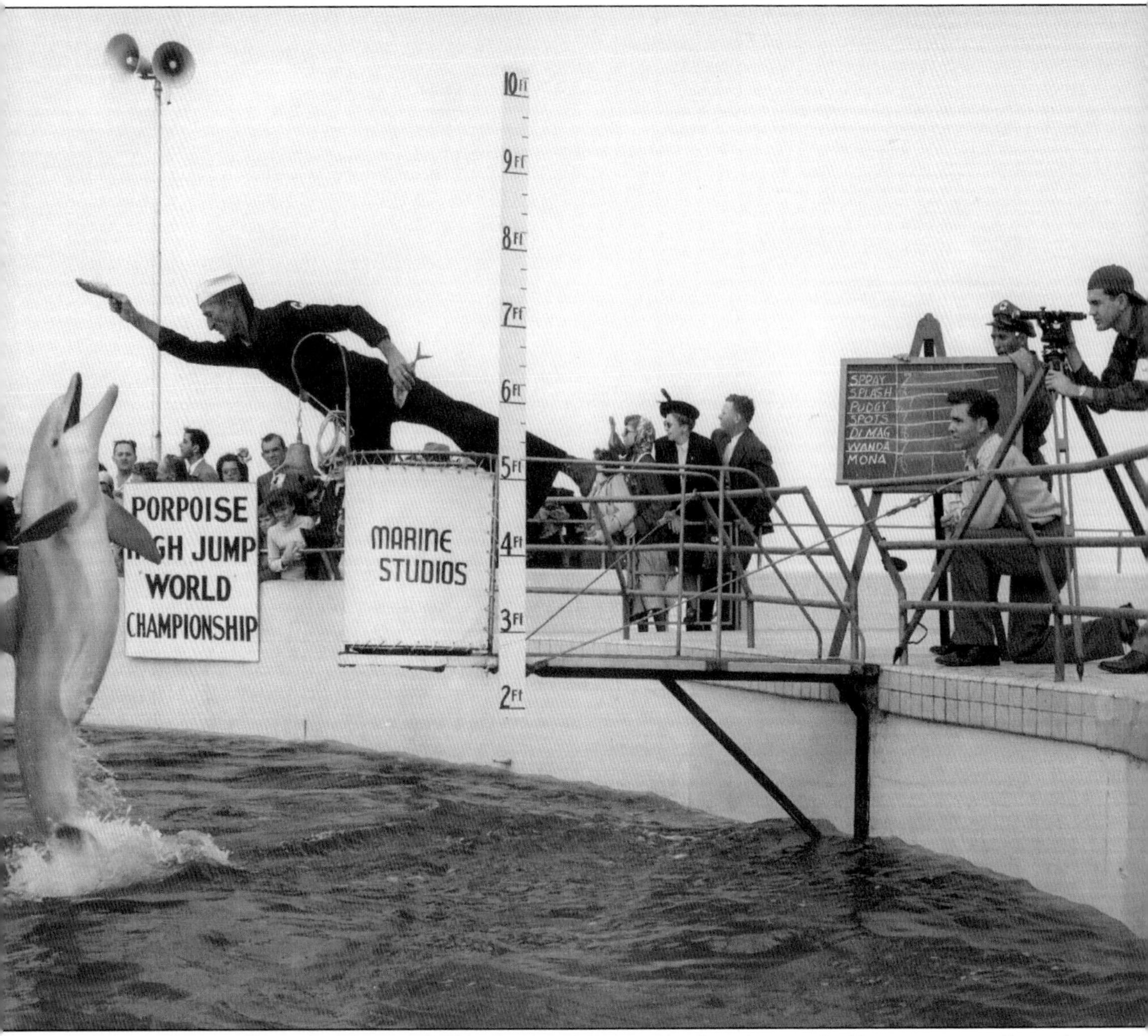

Many topics of interest attracted film crews to Marine Studios, but probably none more than the bottlenose dolphins. With early performances featuring their world-famous jumping abilities, guests frequently wondered how high a dolphin could actually jump. In an effort to answer that question and gain a little additional publicity, Marine Studios created the Porpoise High Jump World Championships. While veteran jumpmaster Mitch Lightsey undoubtedly used his skill to identify each of the jumping dolphins, his colleagues used official surveying tools and a larger-than-life measuring stick to clearly gauge their efforts and a tally board to determine the winner. A closer look reveals that dolphins Mona, DiMag, and Splash were in a three-way tie at the time this photograph was taken, having all reached the eight-foot mark. Guests intently watching this event may have missed another amazing display directly behind them, the impressive feats of wild dolphins swimming and jumping naturally along the bordering coastline.

A Paramount newsreel filmed in 1940 featured the invention of an underwater camera called the bathygraph. The camera was invented by famed deep-sea motion picture photographer Vincent Palmer, who later joined the Marine Studios staff. A real bathygraph was used to film the first live dolphin birth in the circular oceanarium on February 13 of that year, but the camera shown here was just a prop.

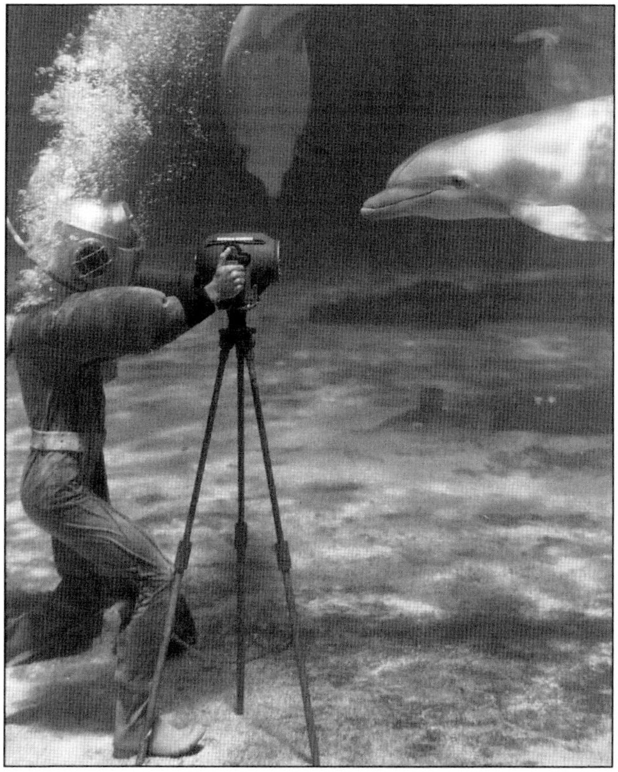

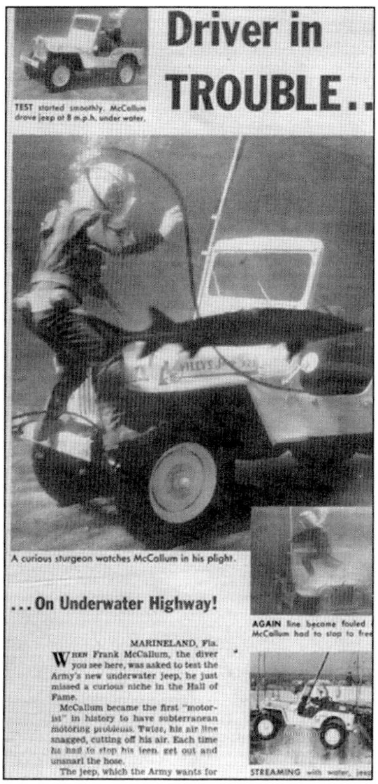

Driver in TROUBLE...

TEST started smoothly. McCallum drove jeep at 8 m.p.h. under water.

A curious sturgeon watches McCallum in his plight.

... On Underwater Highway!

AGAIN line become fouled McCallum had to stop to fre

MARINELAND, Fla.

WHEN Frank McCallum, the diver you see here, was asked to test the Army's new underwater jeep, he just missed a curious niche in the Hall of Fame.

McCallum became the first "motorist" in history to have subterranean motoring problems. Twice, his air line snagged, cutting off his air. Each time he had to stop his jeep, get out and unsnarl the hose.

The jeep, which the Army wants for

STREAMING with water, jee

In January 1951, newsreel crews again converged on Marine Studios, but this time to film the testing of a new Army amphibious jeep, equipped to operate underwater. It was driven back and forth across the sandy bottom of the Marine Studios circular pool by diver Frank McCallum while Army officers, engineers, and cameramen watched. The jeep ran steadily for one full hour before being removed from the pool.

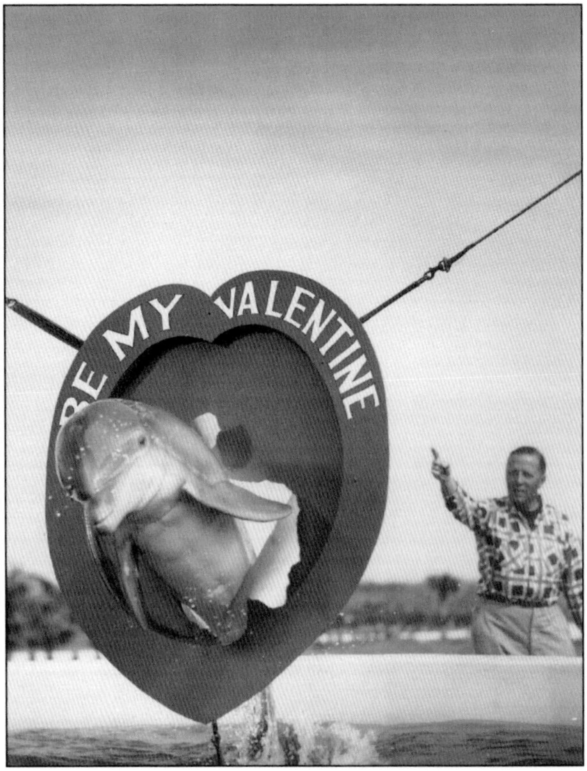

February 1952 marked the one-year anniversary of Flippy's introduction to the press, a fitting time to send valentine wishes to the press and public who had fallen in love with him. Although Flippy's photograph with a Christmas wreath also adorned staff Christmas cards, the celebration of this merry holiday involved much more than a simple image. Year after year, divers constructed coral Christmas trees that were submerged in the pools and decorated with live shrimp, starfish, and other edible ornaments; unwrapped underwater presents that revealed beautiful mermaids; and dressed themselves in Santa suits. Santa even made appearances as the jumpmaster, offering fish goodies to the dolphins. After hours, the staff threw lively holiday parties, where they performed silly skits, traded gag gifts, and enjoyed good-natured celebrating as the close-knit family they had become.

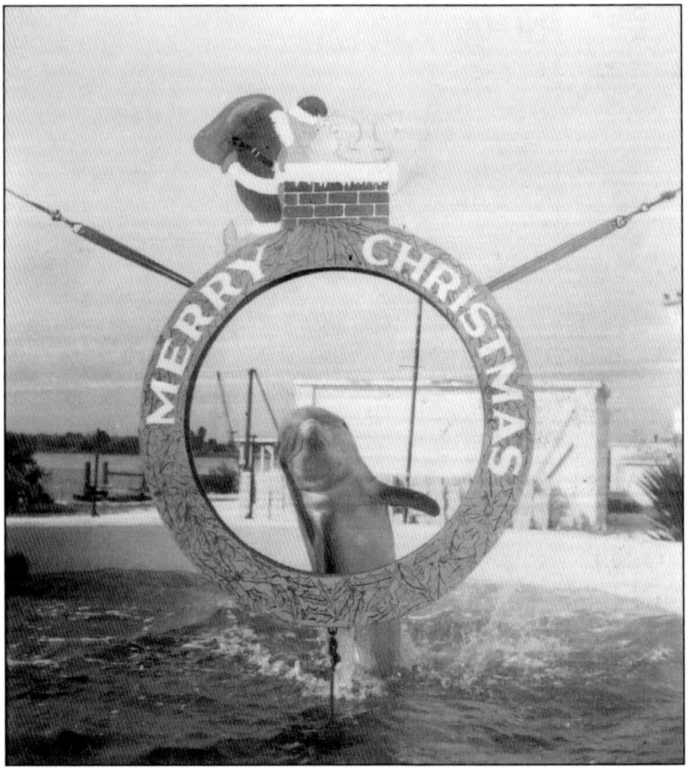

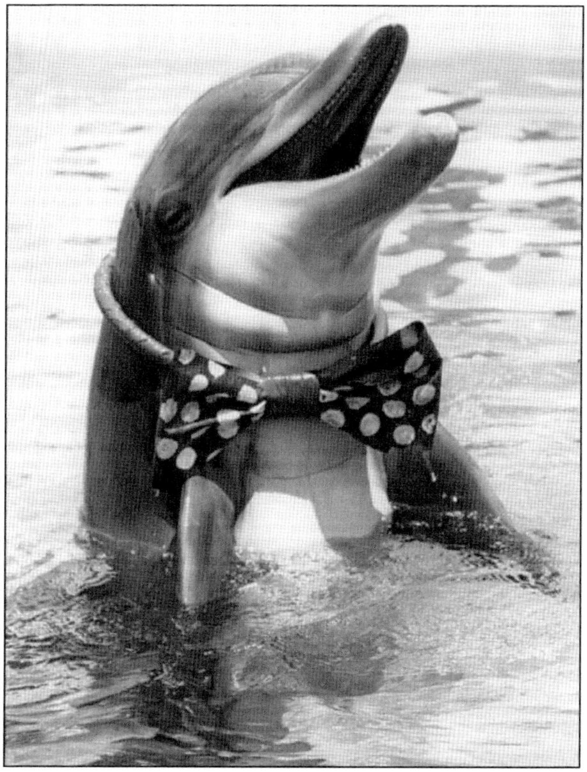

Marine Studios used any opportunity to celebrate, as the caption on this press release image suggests: "Fish To Be Tied: Looking as though he is giving his less well-dressed briny brethren the horse laugh, Flippy the porpoise sports a polka-dotted bow tie with his black and white dress suit. Flippy isn't just showing of. He's a purposeful porpoise and he is calling attention to National Bow Tie Week."

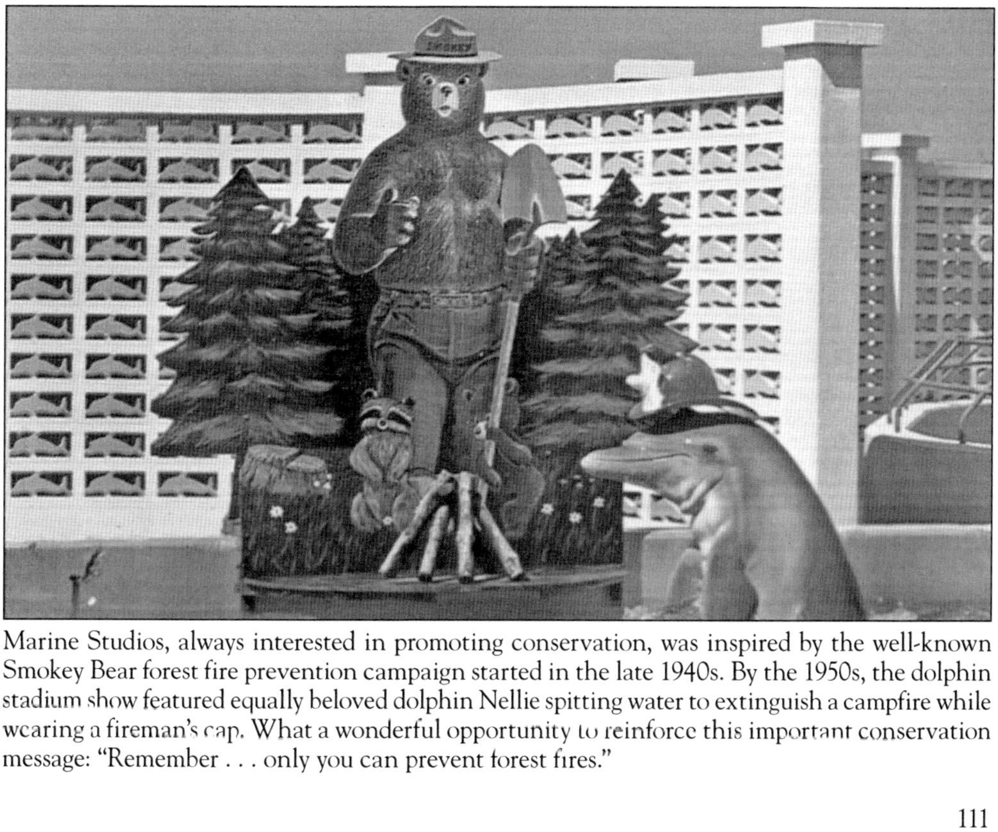

Marine Studios, always interested in promoting conservation, was inspired by the well-known Smokey Bear forest fire prevention campaign started in the late 1940s. By the 1950s, the dolphin stadium show featured equally beloved dolphin Nellie spitting water to extinguish a campfire while wearing a fireman's cap. What a wonderful opportunity to reinforce this important conservation message: "Remember . . . only you can prevent forest fires."

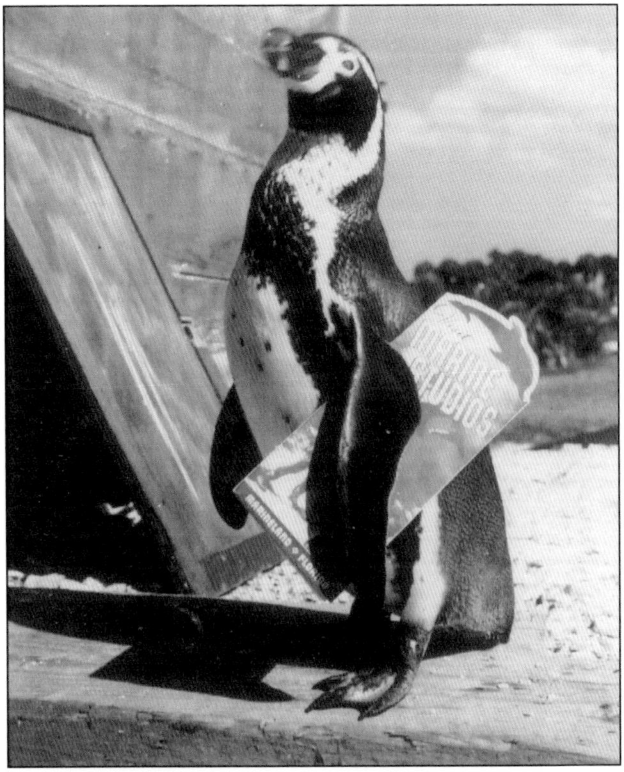

Although always popular for their charisma and ever-present smile, the dolphins were not the only ones to make it in front of a camera for promotional reasons. This African penguin was shown toting his own Marine Studios brochure as part of an advertising campaign to bring in more visitors and perhaps to show them the proper way to behave upon their visit. Meanwhile, fashion photographer Toni Frisell took advantage of the Marine Studios' exotic surroundings to get her best shot, one that did not even need an animal in it, but took a lot of courage to obtain. Frisell was, in fact, known for outdoor fashion photography of active women, including Eleanor Roosevelt and Jacqueline Kennedy.

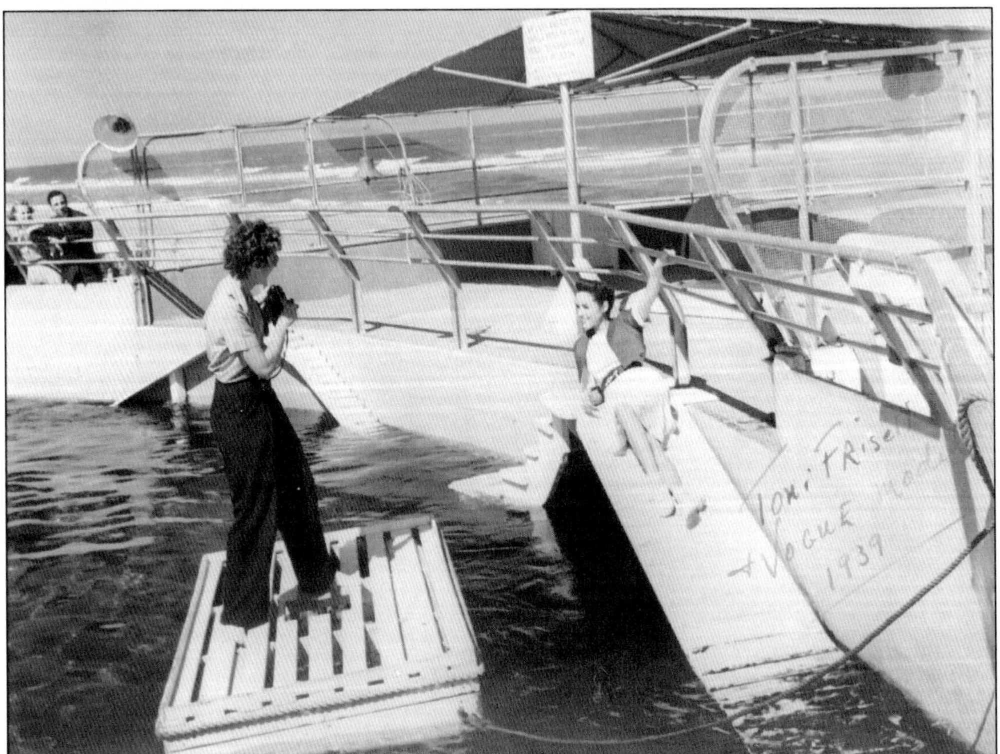

Of course, the original vision for Marine Studios involved the filming of major motion pictures. *Revenge of the Creature*, sequel to *Creature from the Black Lagoon*, was filmed on site for three weeks in July 1954. Shot in the oceanarium during record summer crowds, the facility remained open and visitors received the unexpected bonus of watching a Hollywood movie being made. Accomplished swimmer and diver Ricou Browning was the man inside the creature suit in the underwater shots. Reportedly, the photographers and swimming guards would have to relocate curious sea turtles that would bite large pieces from Browning's sponge-rubber suit and sometimes snap at the actors. No doubt inspired by Marine Studios' dolphins, Browning went on to create the popular *Flipper* movies and television series of the 1960s.

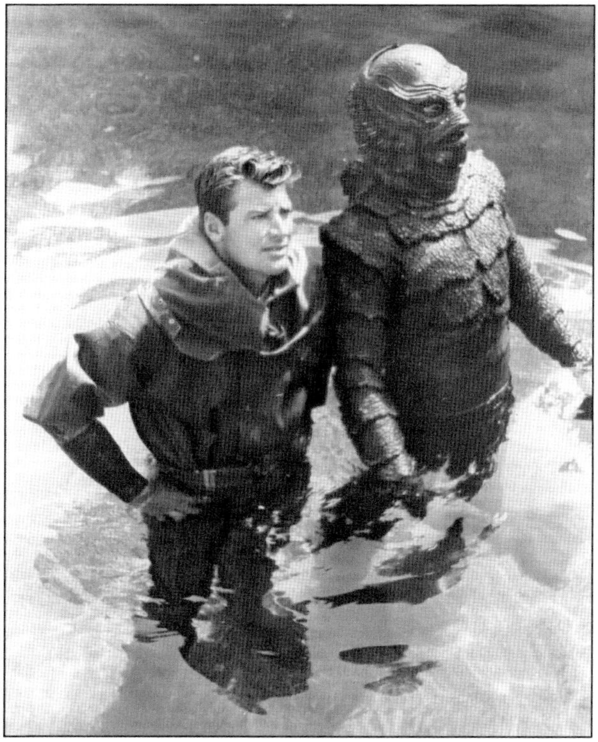

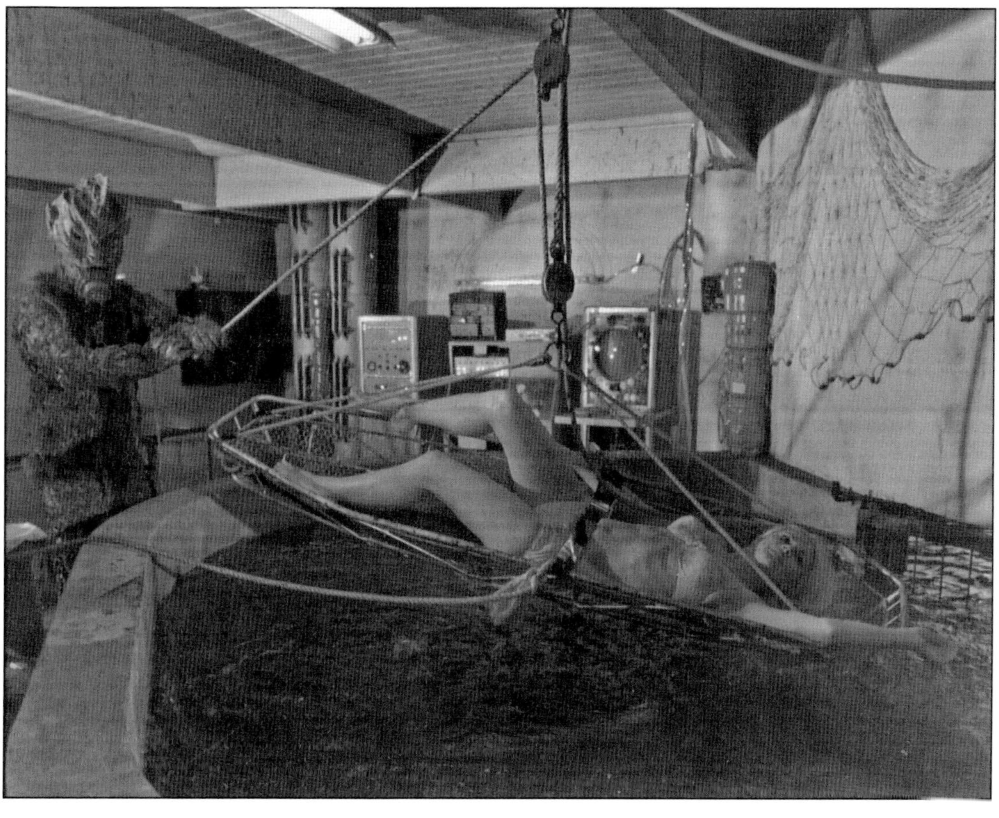

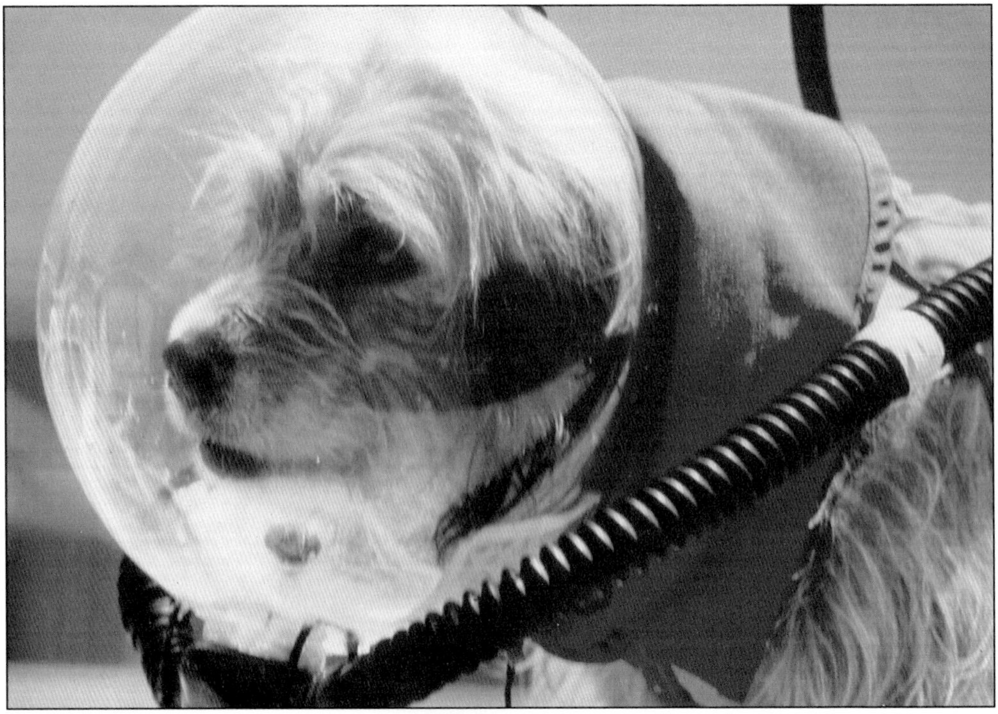

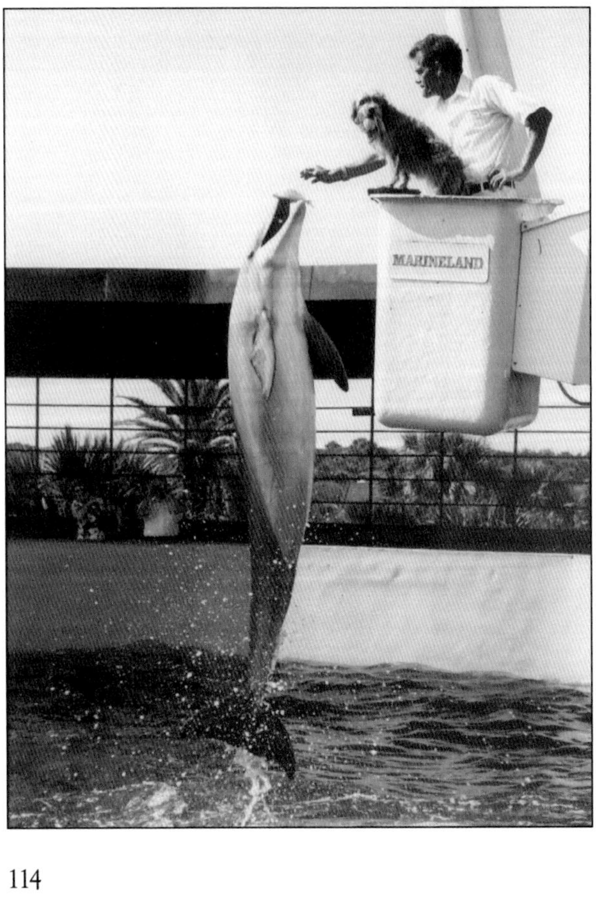

In the early 1980s, *Benji Takes a Dive* was filmed at Marineland. According to trainer Frank Inn, the first challenge was figuring out how to get Benji in the water. Eventually, a special bubble helmet and jacket were made for America's most beloved dog to scuba dive in, and the stage was set for a truly memorable television special.

The training of Benji for this role was challenging. Frank Inn first acclimated Benji to swimming in the customized scuba gear in a backyard swimming pool. Not only did Benji manage to become an accomplished diver, he also performed with the dolphins as a jumpmaster, holding still while a dolphin leapt into the air and snagged a dead fish straight from his mouth.

Seven

DEVELOPERS

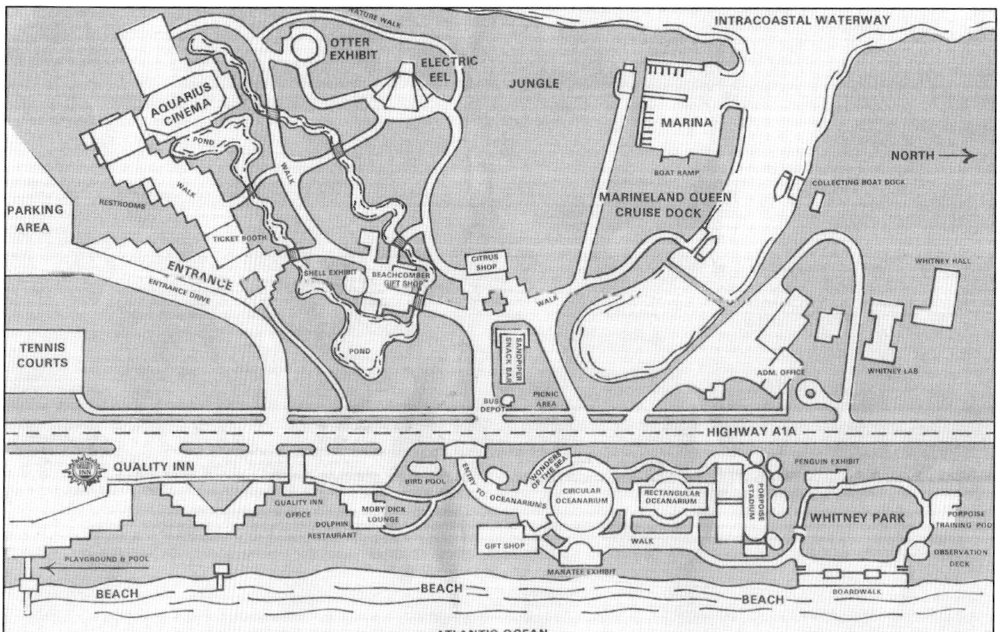

As time moved forward, Marine Studios obtained a new name and a new logo, reinventing itself as Marineland of Florida. Expanding onto property on the west side of Highway A1A in the 1970s, new and interesting exhibits and animals were added. In the subsequent decades, following trends in science and public demand, even more changes occurred, and to this day, the ever-evolving facility continues to move forward, while still managing to fondly hold on to its historic roots.

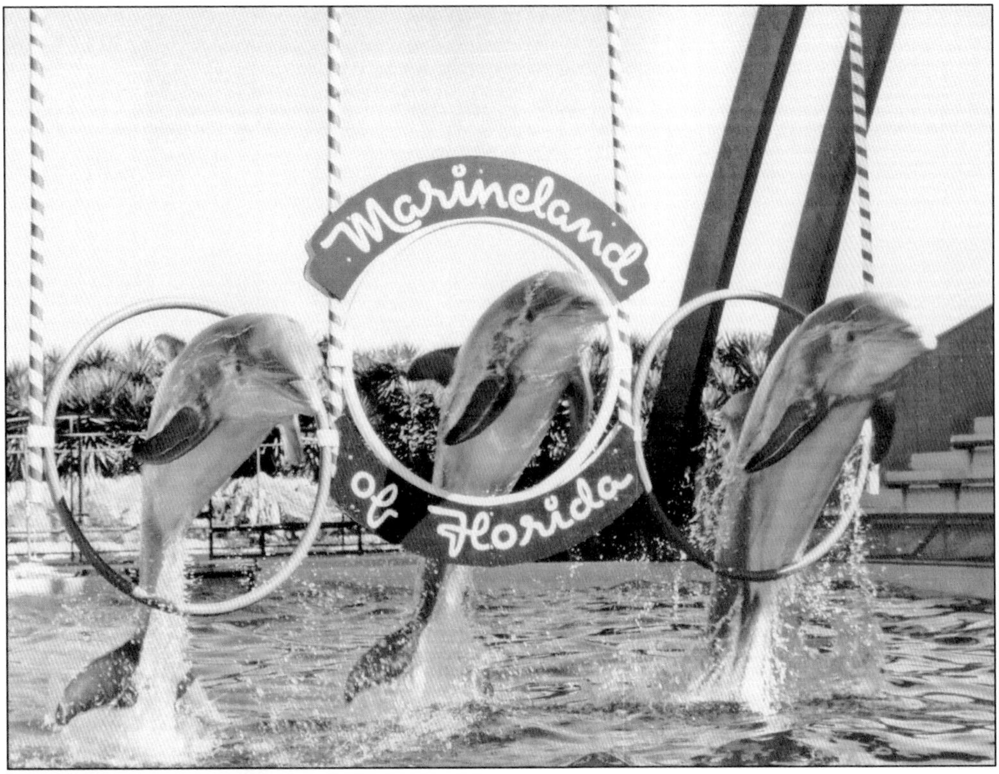

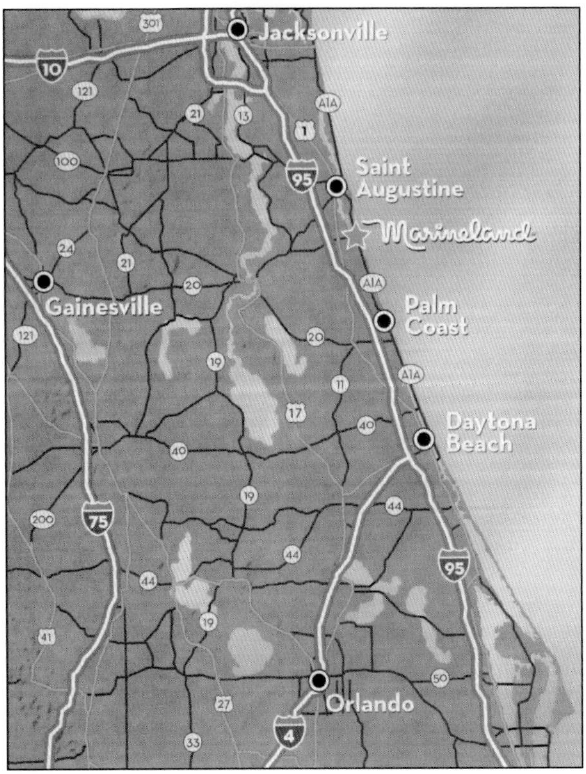

Marine Studios' mailing address had been in the town of Marineland, Florida, ever since Marineland incorporated in 1946. Understandably, this caused some confusion, and the names were sometimes used interchangeably. Since Marineland was an authentic, albeit small, city with its own post office, fire department, restaurants, and service stations, the town was also clearly marked on any map of Florida. It was a brilliant marketing move when Marine Studios changed its name to Marineland in 1961 and was immediately, and literally, on the map. The relationship between the town and the oceanarium was mutually beneficial; the town's first mayor, Milton E. Bacon Jr., was the director of public relations at Marine Studios, and the town itself was founded by Marine Studios' founder C.V. Whitney.

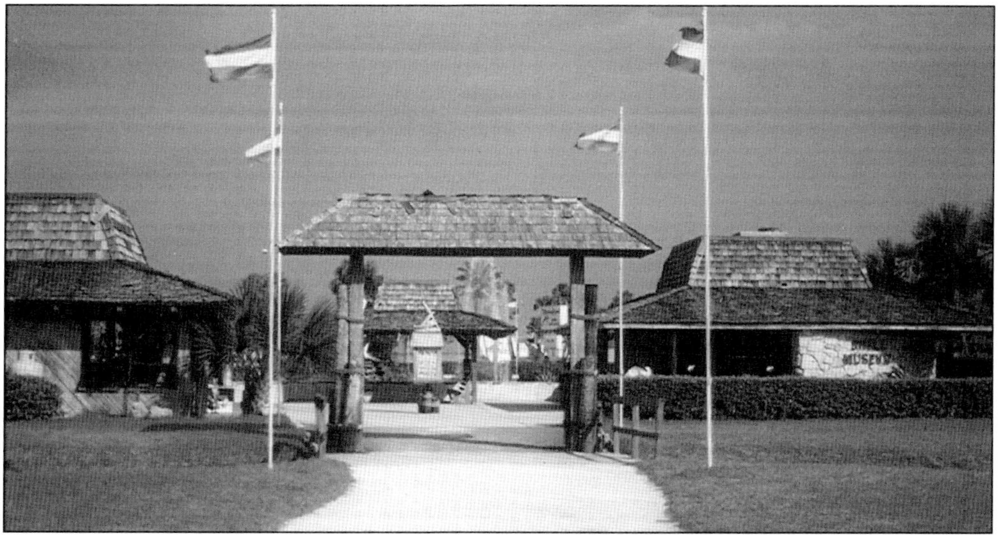

Landscaped walkways led guests to the new Entrance Plaza (above) in the 1970s. After a short stroll along a nature path, introducing the visitors to the subtropical Florida jungle, visitors would arrive at the state-of-the-art Aquarius Theater, which was designed after the Academy Awards screening theater in Hollywood. Here, Space Vision, the latest in film technology, was displayed as guests reached into the air from their seats, grabbing for the three-dimensional images floating from the screen. Titled *Sea Dream* and filmed at Marineland in 1978, the movie allowed guests yet another way to be fully immersed in the oceans without ever really getting wet. Forty years after its opening, Marineland was still bringing the ocean to people who would otherwise never know the magic it contained.

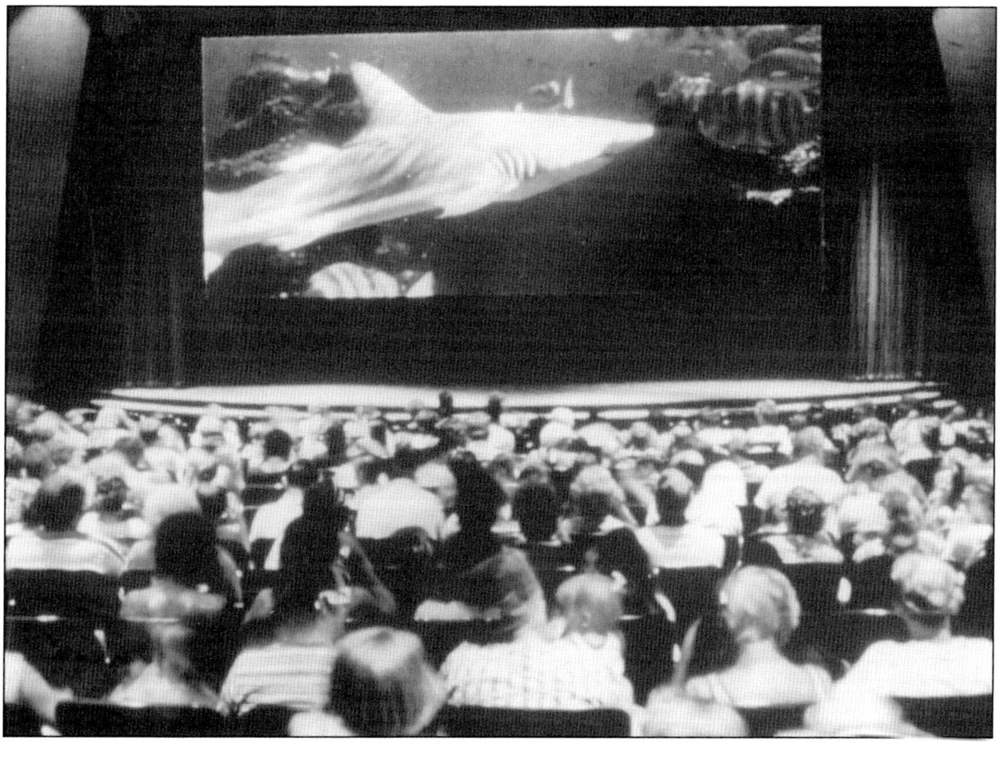

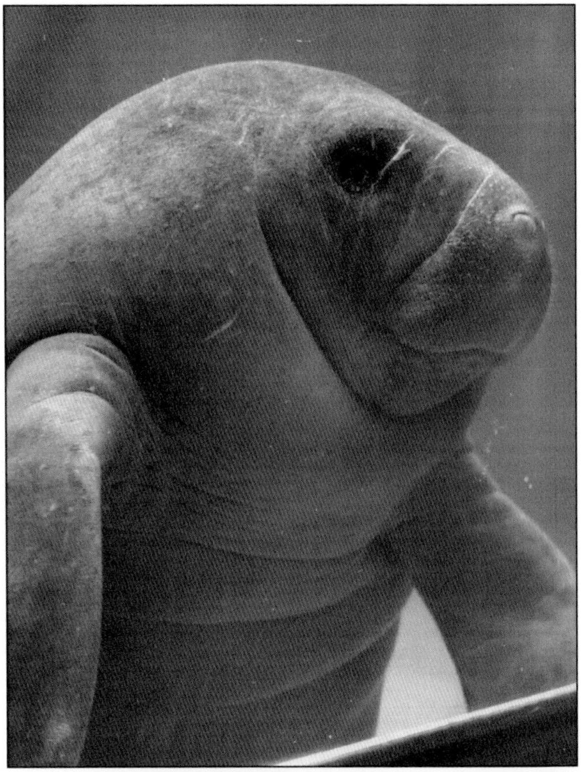

In the late 1960s, the marine mammal collection expanded with the addition of manatees and Amazon river dolphins, who were housed together in the same warm-water exhibit, adjacent to the original circular oceanarium. Since neither species is known for its great athletic prowess, this display was designed more for quiet observation, a nice contrast to the high-energy shows featuring bottlenose dolphins. Both manatees and river dolphins were rather unknown at the time, and extensive discoveries regarding their behavior and adaptations emerged from their stay at Marineland. In fact, on June 15, 1971, Marineland welcomed the birth of the first live river dolphin in a zoological setting, marking yet another first for the world's first oceanarium.

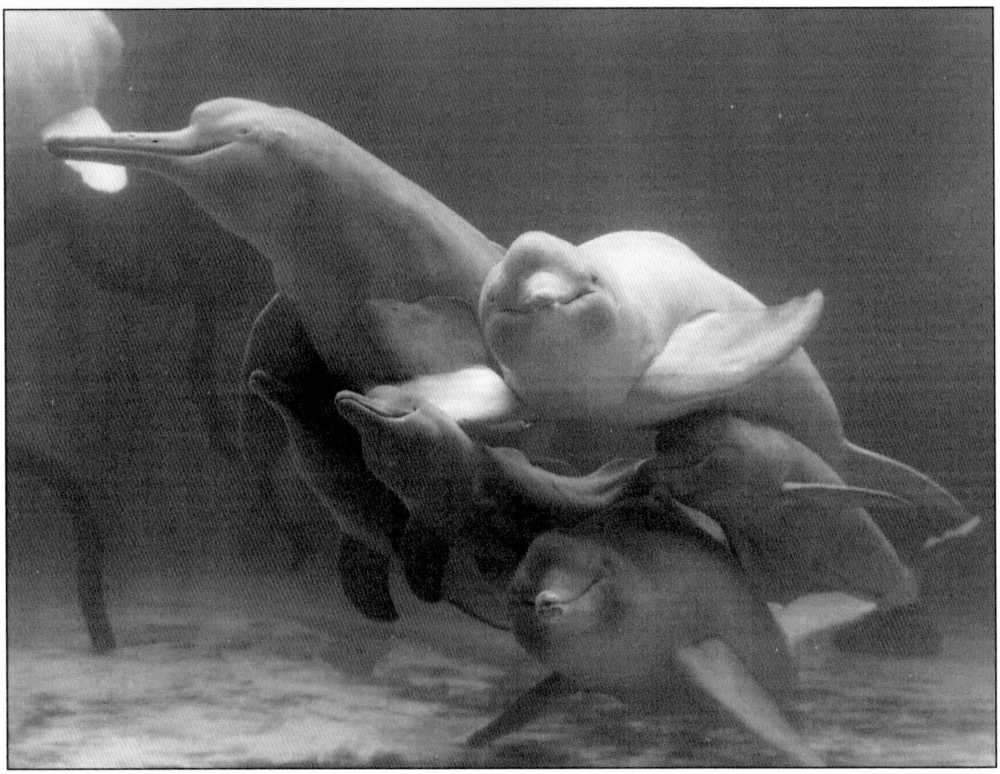

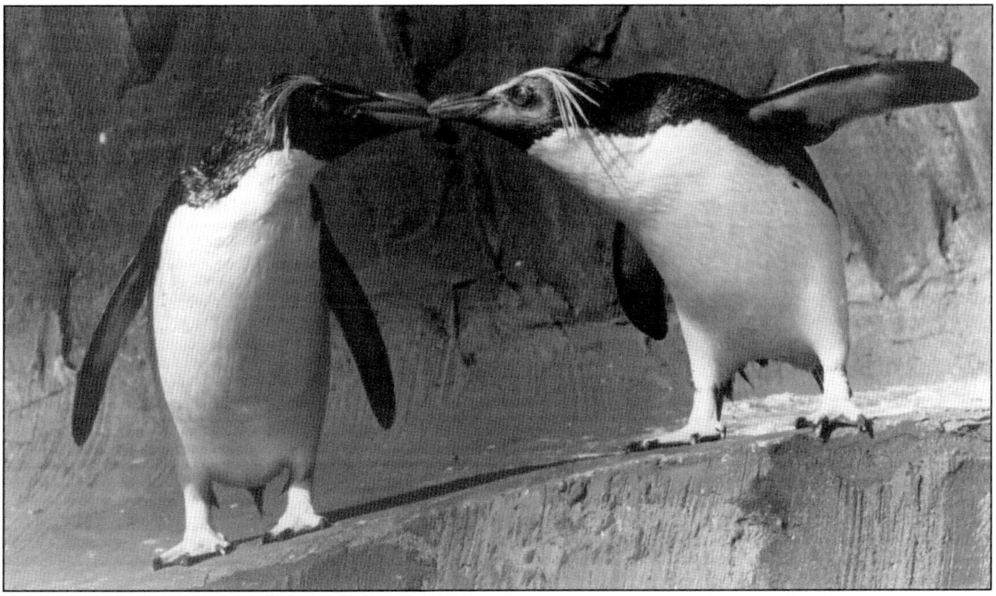

A variety of other animals were added to the Marineland collection in the 1970s. Humboldt and black-footed penguins were exhibited for the public's observation, while California sea lions performed in a new and popular show. All of these animals made their home in Whitney Park, an addition extending just north of the dolphin stadium and named after original founder Cornelius Vanderbilt Whitney. A raised platform along the east side of Whitney Park provided guests an excellent vantage point from which to study the Atlantic Ocean and the Matanzas Inlet. This boardwalk was, both literally and figuratively, a bridge for visitors to learn about the animals in the Marineland exhibits and then gaze out into their wild environments.

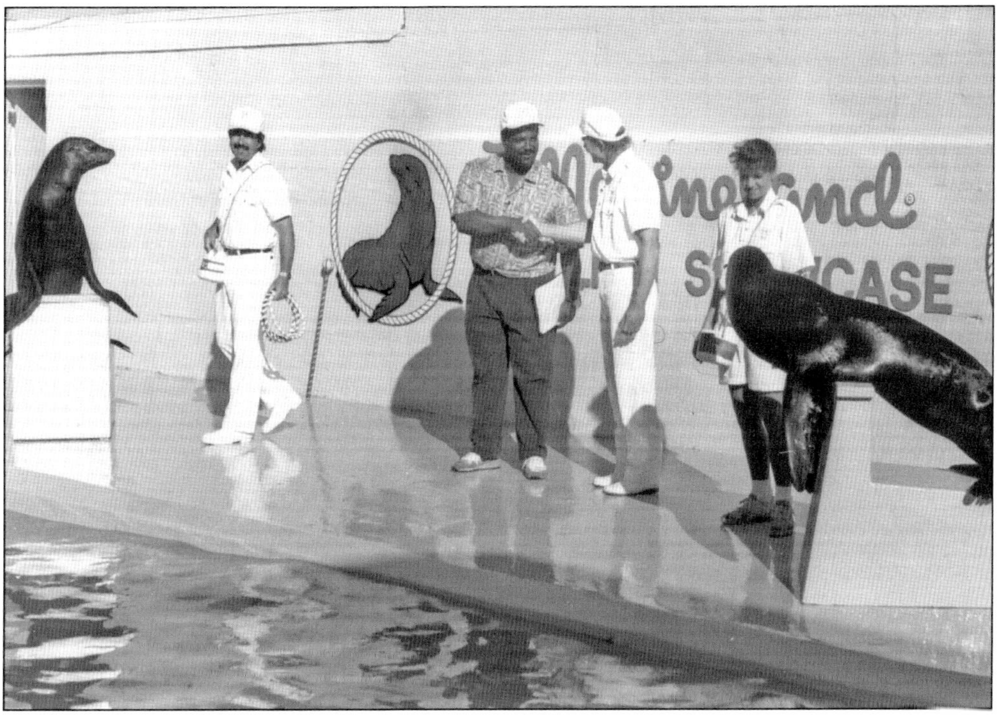

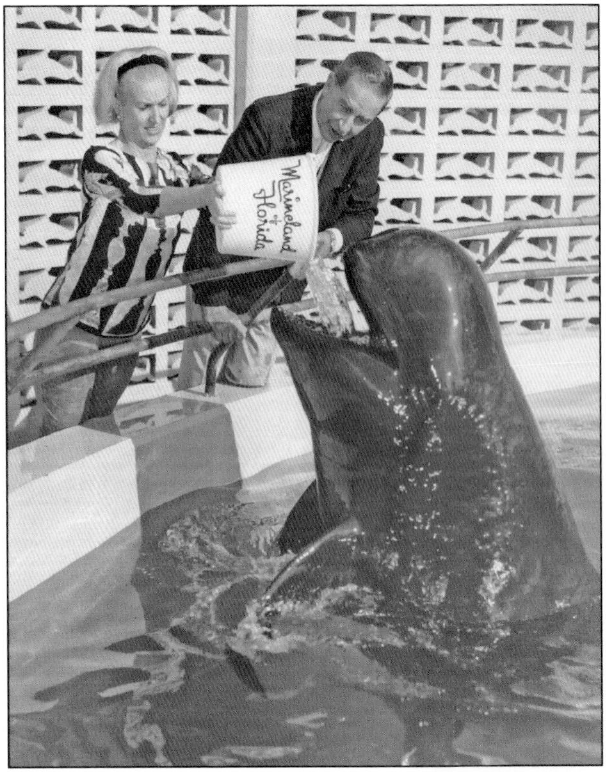

The Whitneys' involvement in Marineland remained strong throughout the years. They were instrumental in park improvements of all types, from expanding the animal collection to the details of its unique architecture, including the creation of the one-of-a-kind dolphin bricks seen in this picture and still present at the facility today.

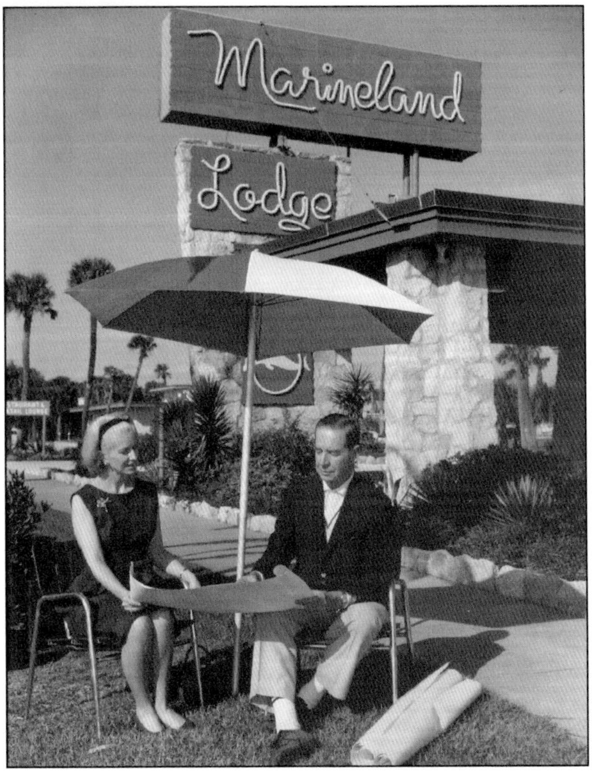

The affiliation between the Whitneys and Marine Studios was the longest of all the original founders. Douglas Burden had retired from active participation in the late 1960s, and Ilia Tolstoy passed away in 1970, leaving Whitney as the lone active remaining founding member. He and his wife, Marylou, were frequently seen around the oceanarium and throughout the town.

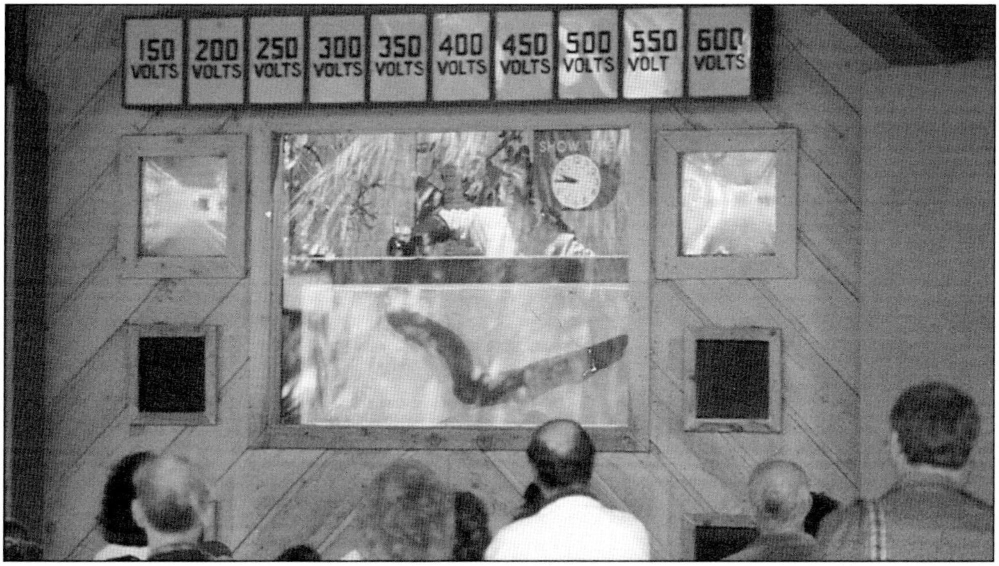

Electric eels and other marine fish were displayed in smaller exhibits, allowing them to be highlighted rather than disappearing into the depths of an enormous rectangular oceanarium. In the 1970s, the electric eel put on a show of its own in its custom-designed pavilion that used audiovisual equipment to display the 600 volts of electricity the animal was capable of generating. Located adjacent to the circular oceanarium in 1957, Wonders of the Sea was a collection of smaller aquariums, which showcased fish and crustaceans, both native and nonnative, and displayed fascinating examples of camouflage and other marine adaptations. Detailed graphics ensured that guests were as educated as they were entertained.

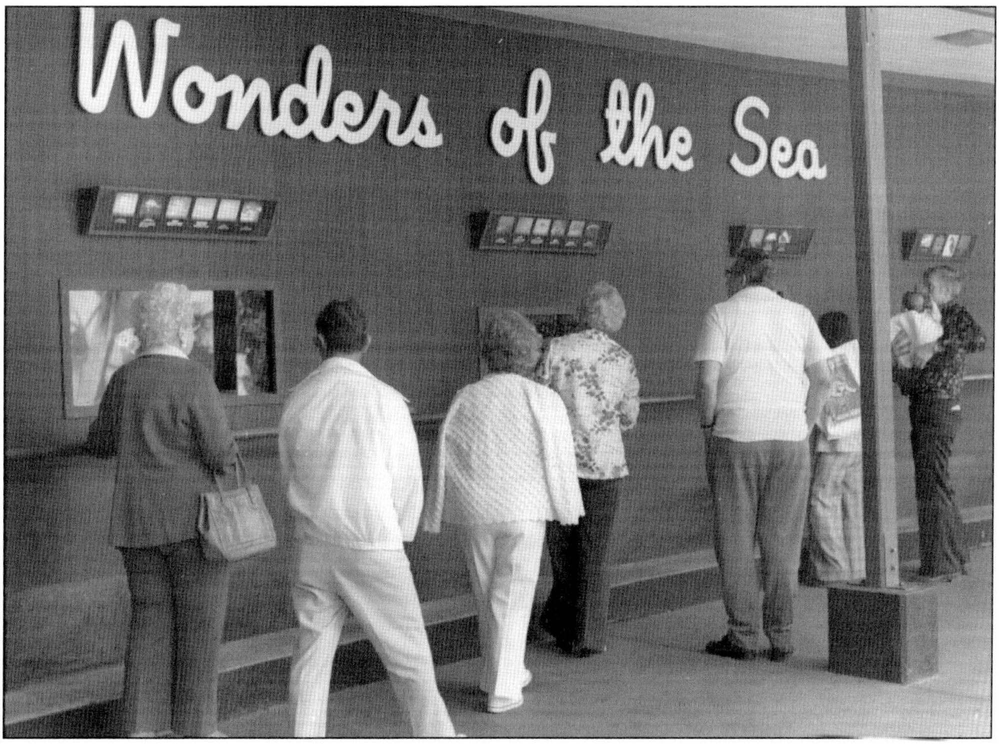

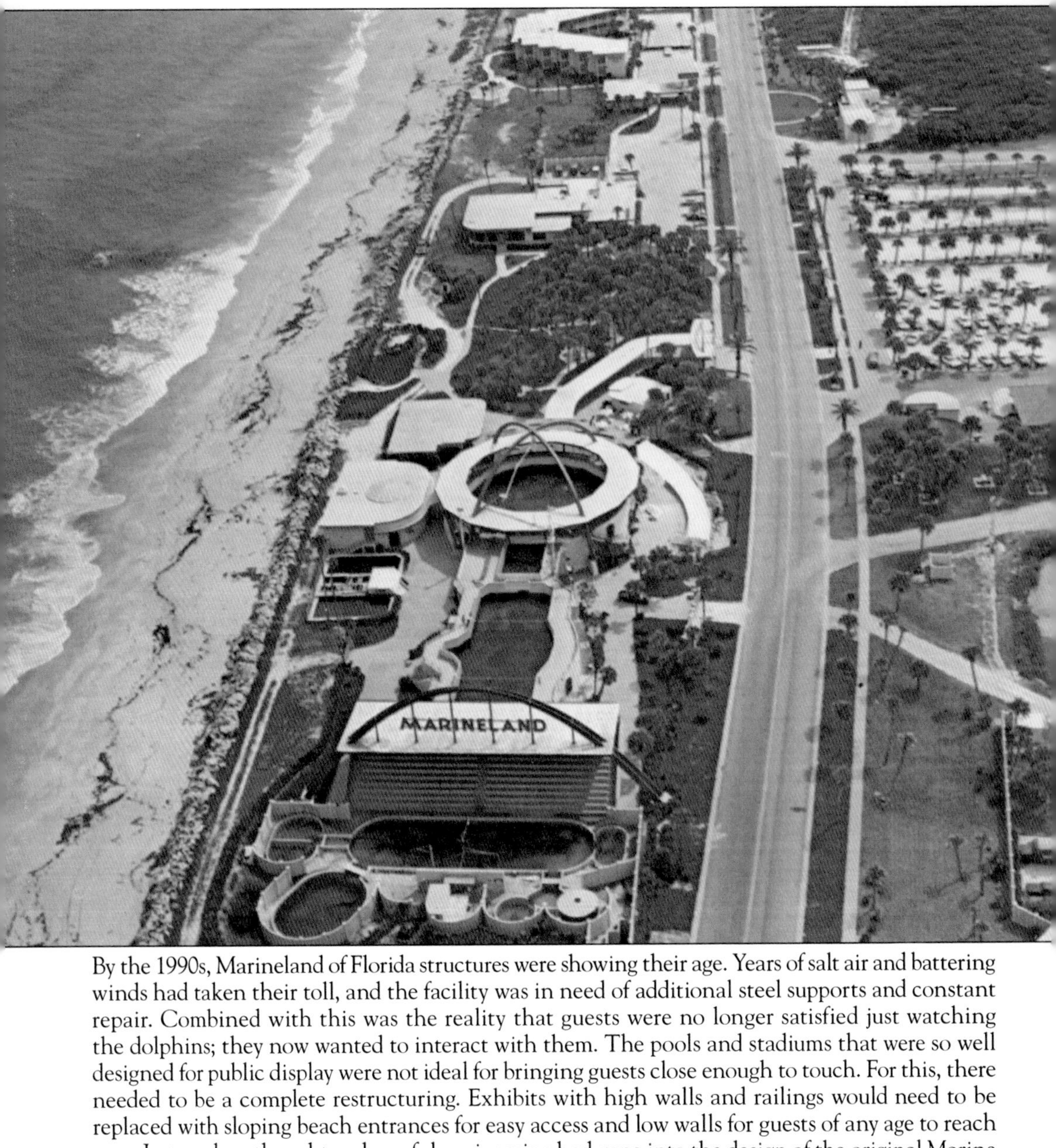

By the 1990s, Marineland of Florida structures were showing their age. Years of salt air and battering winds had taken their toll, and the facility was in need of additional steel supports and constant repair. Combined with this was the reality that guests were no longer satisfied just watching the dolphins; they now wanted to interact with them. The pools and stadiums that were so well designed for public display were not ideal for bringing guests close enough to touch. For this, there needed to be a complete restructuring. Exhibits with high walls and railings would need to be replaced with sloping beach entrances for easy access and low walls for guests of any age to reach over. Just as close thought and careful engineering had gone into the design of the original Marine Studios, this new facility would also require nothing but the best. A team of designers, engineers, architects, and animal care experts were prepared for just such a task. With the acknowledgment of the end of one era and excitement for the beginning of a new one, Marineland, born Marine Studios, became Marineland's Dolphin Conservation Center.

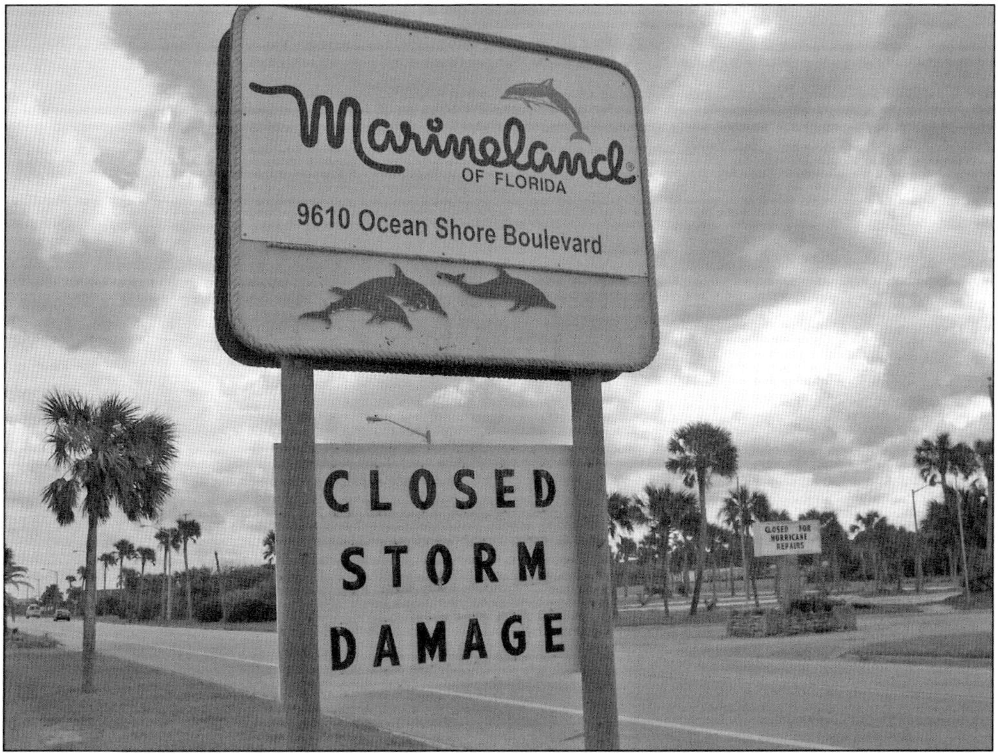

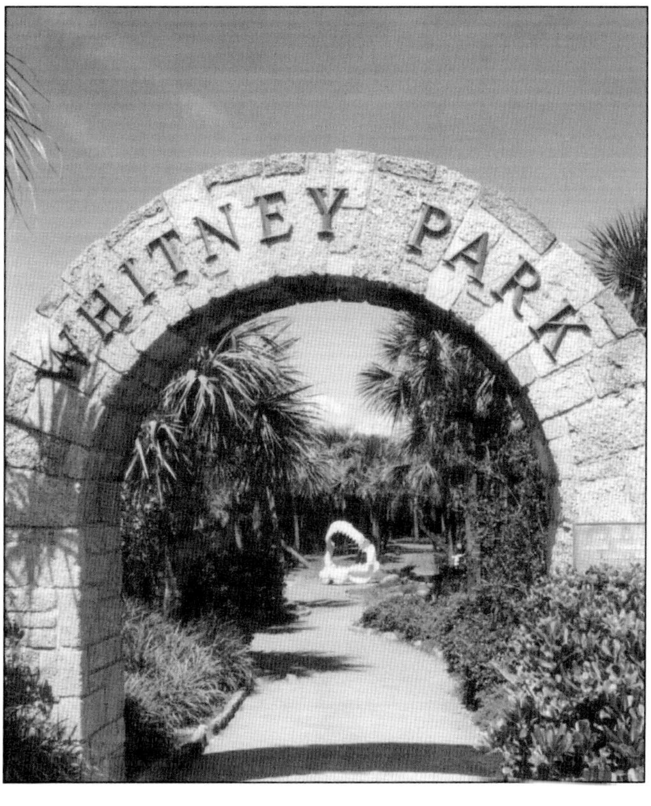

Amid the planning and design of the new facility, standard operation went on as usual, with daily shows, presentations, and animal exhibits. That is until the summer of 2004, when three hurricanes tracked right over Marineland. Although animals and staff were fine, the oceanarium was not. Marineland closed for the final time in September 2004 and relocated all of the animals living there, except the dolphins, which remained safe in their original habitats; however, efforts to build their new home were redoubled. Just north of the dolphin stadium in the parcel of land know as Whitney Park, piping was laid and cement was poured for the new Dolphin Conservation Center, a 1.3 million-gallon dolphin habitat.

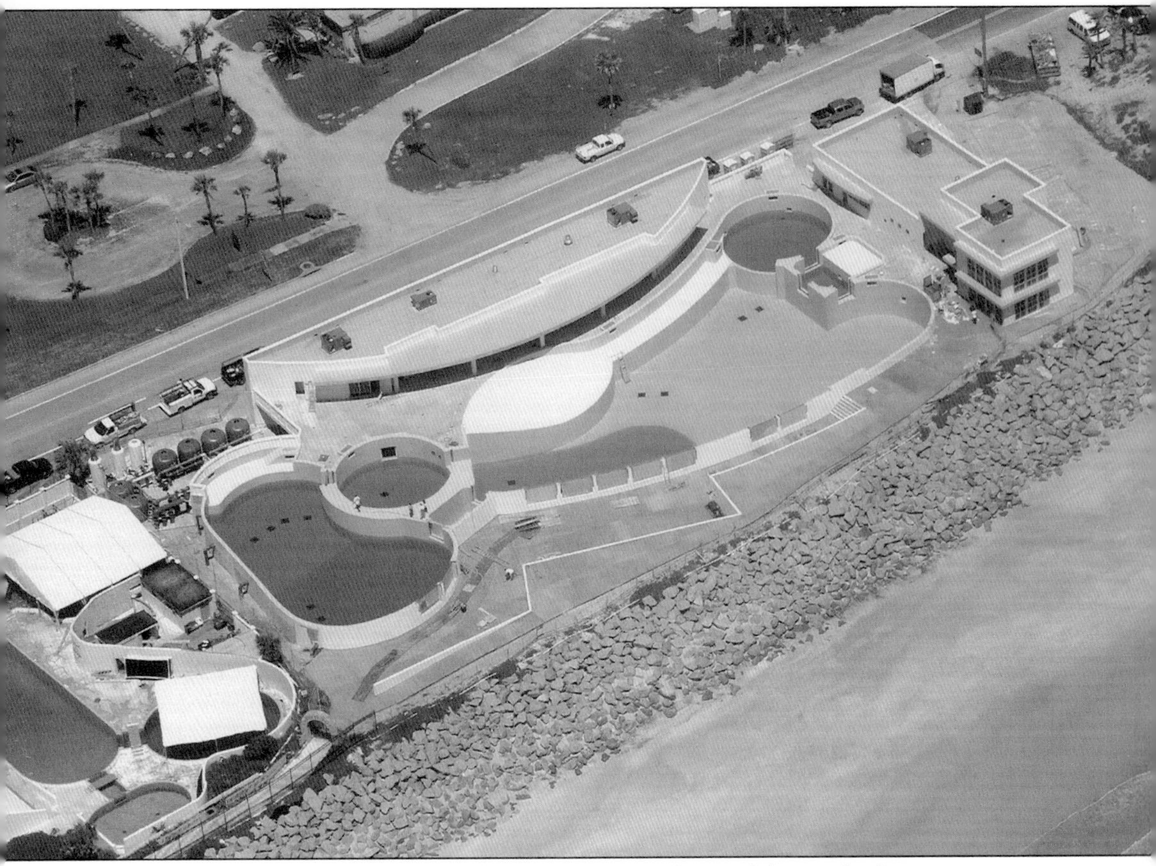

Slowly but surely, the new structure began to emerge. State-of-the-art facilities, viewing areas, and easily accessible habitats, in addition to a medical lab, defined the project. With a nod to its Marine Studios roots, Marineland's Dolphin Conservation Center adopted the nautical moderne style of architecture and included facilities for researchers and filmmakers in its plans. One by one, the dolphins were transported by stretcher from the original Marineland dolphin stadium into their new home. On a cold winter day in early 2006, the Dolphin Conservation Center's doors opened to the public, and in a grand opening reminiscent of that great day in 1938, traffic jams blocked Highway A1A for miles in each direction as guests visited in record numbers. In two days, 20,000 people attended the festivities. The grand opening ceremony incorporated family members of Marine Studios' original founders, including C.V. Whitney's widow, Marylou Whitney; W. Douglas Burden's daughter, Wendy Burden Morgan; and Arthur McBride's niece, Cathi Haggerty.

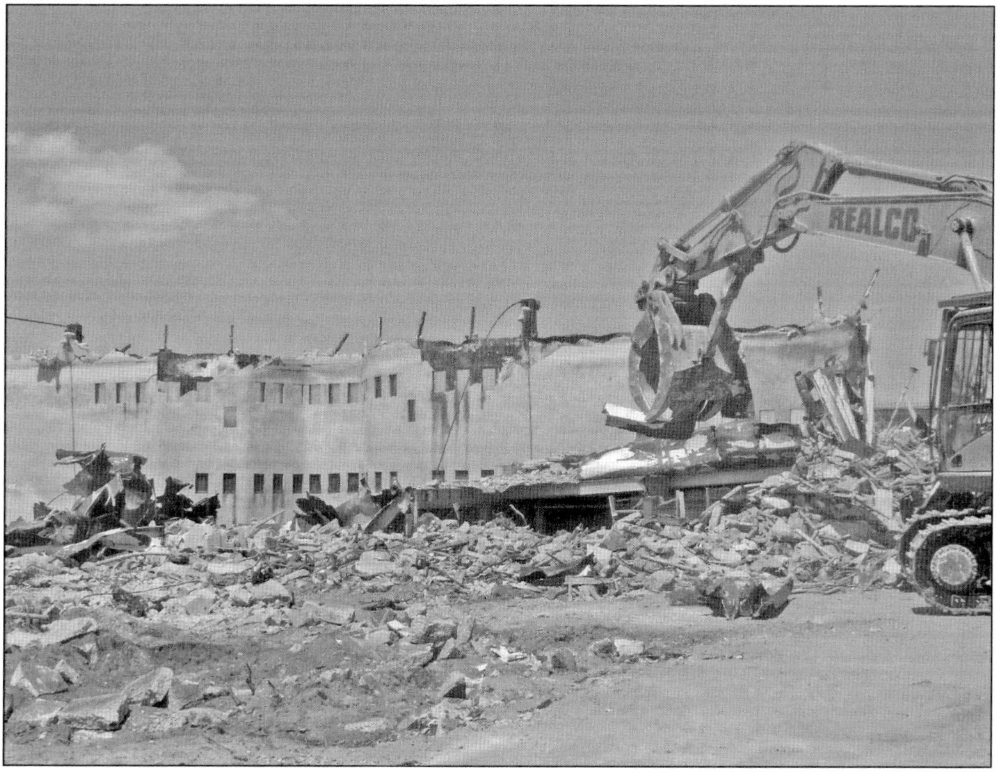

With a successful grand opening behind them, the staff of Marineland's Dolphin Conservation Center envisioned ways to honor the original structures of Marine Studios. While no longer needed to hold water or animals, the circular and rectangular oceanariums would be ideal for use as classrooms, museums, or meeting halls. Structural engineers were brought in to assess the condition of the buildings and recommend repairs. To everyone's great disappointment, the reports came back requiring such extreme retrofitting and restructuring that the repaired buildings would bear no resemblance at all to the originals. With a heavy heart, the almost 70-year-old structures were razed.

DOLPHIN CONSERVATION CENTER
Marineland

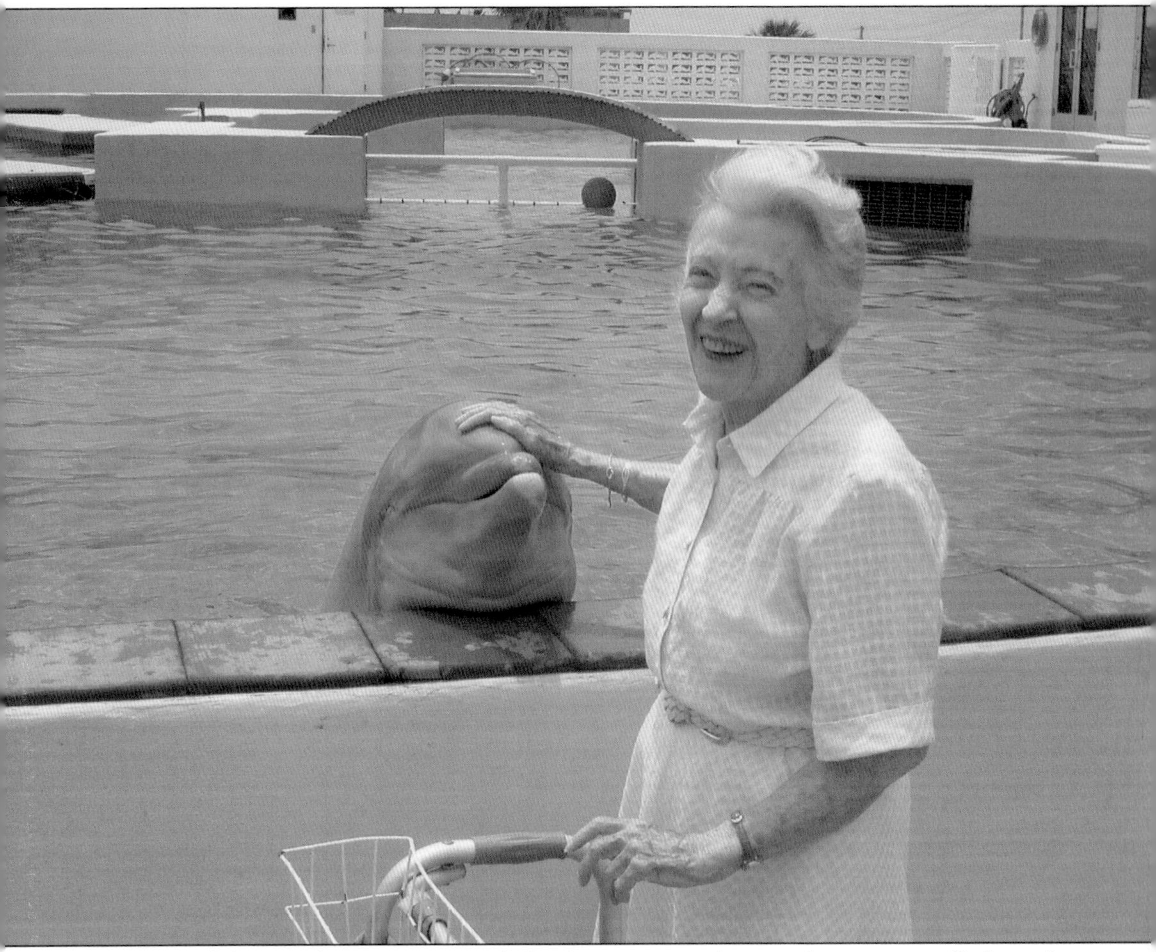

As news of the grand reopening spread, St. Augustine residents with stories to tell about their connections to Marine Studios began to visit the new facility. Whether they had worked at the gift shop during their teen years or spent their honeymoon visiting the oceanarium 70 years prior, it seemed everyone had a tie to the local landmark. With such success in the care of bottlenose dolphins, many of the dolphins from the 1960s and 1970s were still living at Marineland's Dolphin Conservation Center, and retired animal trainers came out to see their old friends. Shown here is Betty McBride, widow of Marine Studios' first curator, Arthur McBride, who died so young but made so many lasting discoveries and held such a high regard for true science. Betty McBride is giving Lilly, the original blonde dolphin collected in 1964, a pat on the head as two old friends are reunited.

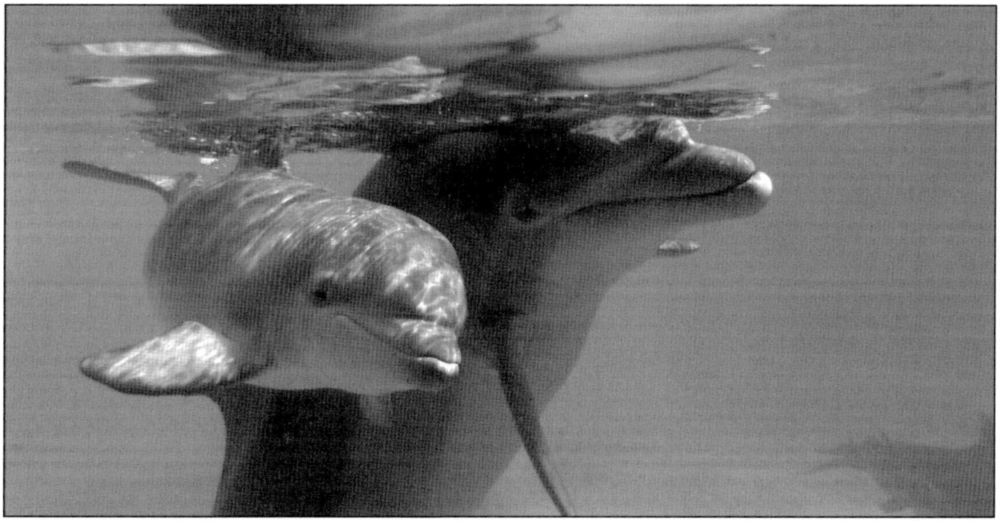

As a successful dolphin interactive facility offering once-in-a-lifetime opportunities for many guests, Marineland's Dolphin Conservation Center has not forgotten the tenets on which Marine Studios was founded. Successful breeding continues to be a focal point, as mother Roxy, born in the original facility, is shown swimming above with her new calf Tocoi, born in the new dolphin habitats. Groundbreaking research is also an integral part of the facility's mission, as seen below in the hearing test being given to Chubby, a Marineland resident since 1972. Photographers and filmmakers continue to take advantage of the viewing opportunities, while guests continue to be exposed to a world they would not otherwise know. Marineland's Dolphin Conservation Center, built with the needs of the animals and the visiting public in mind, continues to thrive with a visionary eye toward the future and a proud connection to its celebrated past.

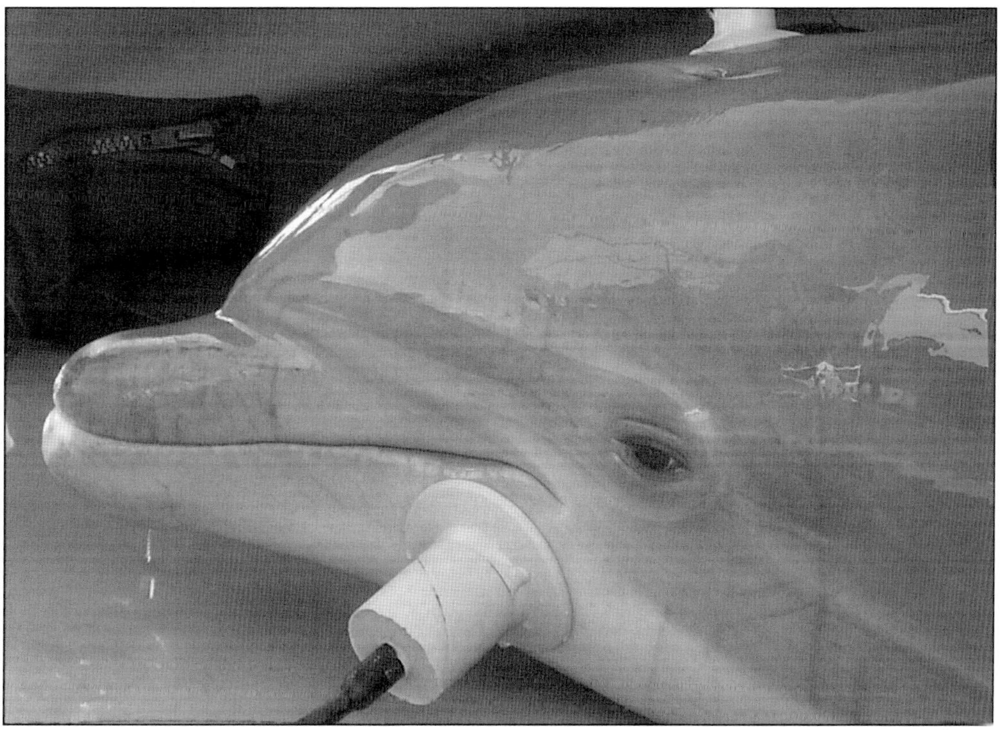